Modern Extension System *and* Participatory Extension Approaches for Planning

NIPA® GENX ELECTRONIC RESOURCES & SOLUTIONS P. LTD.
New Delhi-110 034

About the Authors

Sri Bidyadhar Maharana (1944), a postgraduate in Agronomy from Odisha University of Agriculture & Technology (OUAT), has about 55 years of government service in the Agriculture Department of Odisha and Consultancy experience. As a Certified trainer by the Department of Personnel and Training, Government of India, he trained many high-level and middle-level officers on NRM, Watershed Management, Project Planning, Micro-planning, Participatory Rural Appraisal, Extension Management, and different subject matters for about 20 years in IMAGE and ICAR Institutes. He was a key district-level resource person during the implementation of T & V System of Extension in Odisha from the year 1977. When NATP was pilot-tested in seven states of India in 1998, he took the lead in the preparation of the Strategic Research & Extension Plan (SREP) in four districts of Odisha with the support of a multi-disciplinary team. He was a National Professional in a GOI-UNDP-assisted Sub-programme "Strengthening NRM and Sustainable Livelihoods of Women" in Tribal Odisha for five years. As a freelancer Consultant, he has worked in various organizations like the Asian Development Bank, UNDP, UNICEF, CARE, India, International Water Management Institute (IWMI), and DFID-supported projects for about 15 years. He was a National Level Monitor (NLM) for the Ministry of Rural Development, Government of India, for monitoring all Rural Development Schemes in Jharkhand, West Bengal, and Nagaland for three years. He has provided consultancy to reputed NGOs like Agragamee, XIM-CENDERET, CYSD, PRAGATI, NIRMAN, IGGSS, OPSL, RICOR, LAVS, etc. for many years till now. He has conducted several studies for APICOL, OPSL, RICOR, IWMI, CARE, India, SAMBODHI, OTELP, WORLP, NABARD, etc. He has developed user manuals on Land and Water Management for OWDM, OTELP, Decentralized GP planning for UNICEF, Women and friendly GP Plan for State Institute of Rural Development (SIRD), and RKVY manual for the Government of Odisha and prepared many projects for government. He has written many technical booklets for UNDP and the Nutritional Garden manual for IGGSS. He was associated with IWMI, Srilanka, to prepare a Climate Smart Village Plan for Odisha. He has participated in Doordarshan and All India Radio (AIR) as a resource person for about 20 years. He has authored two Books on "Sustainable Agriculture: Theory & Practice" in association with Professor (Dr) S.K. Sahu and "Climate Smart Agriculture & Food Security" in association with Professor (Dr) B.C. Nayak.

Dr. Chakradhar Satapathy (1942) graduated from the Orissa University of Agriculture and Technology in the year 1966 and completed M.Sc. (ag) in Extension Education being a gold medalist in 1969. He obtained a Ph.D. from Indian Agriculture Research Institute in the year 1981 with highest grade point. After serving for 32 years in Extension Education Stream in Odisha University pf Agriculture & Technology (OUAT), he retired as Professor and Dean of Extension Education in 2002. Thereafter, he served as Vice Chancellor at Kalinga University, Chhattisgarh, Raipur, for a short period of one year. Currently, he is working as Director at Amity Humanity Foundation, Bhubaneswar, under the umbrella of Amity University, Noida, since 2004. He has guided 30 Ph.D scholars besides many M.Sc. students. In addition to more than 200 research papers published in different Journals of research, he has published 22 books both professional and multidimensional in nature He is also the President of the Odisha Society of Extension Education. He has shared experience as a well-known Extension Professional for the benefit of the extension fraternity, scientists, researchers, and policymakers.

NIPA® GENX ELECTRONIC RESOURCES & SOLUTIONS P. LTD.

101,103, Vikas Surya Plaza, CU Block
L.S.C.Market, Pitam Pura, New Delhi-110 034
Ph : +91 11 4386 0225, 9717133558, 9540816132
E-mail: newindiapublishingagency@gmail.com
Website: www. niparesources.com

Print ISBN: 978-93-58876-42-0
ebook ISBN: 978-93-58873-85-6

Composed and Designed by NIPA®.

ଡଃ. ଅରବିନ୍ଦ କୁମାର ପାଢ଼ୀ
Dr. Arabinda Kumar Padhee, IAS
Principal Secretary

Government of Odisha
Department of Agriculture
and Farmer's Empowerment

FOREWORD

Extension Education is one of the prominent branches of agriculture science which has expanded rapidly during the last four decades after implementation of T&V System of Extension and has assumed new dimensions with respect to its different disciplines. The field of 'extension' now encompasses a wider range. of communication and learning activities organized for rural people by educators. Agricultural Extension is a policy instrument that a government can use for stimulation of agricultural development as it provides an effective linkage between farmers, extension workers and agricultural research. I am pleased to know that a reference book has been authored by Mr Bidyadhar Maharana and Dr. Chakradhar Satapathy based on their experience in extension service, teaching and research for five decades. The reference book "Modern Extension System and Participatory Extension Approaches for Planning" not only provides the basics of Extension Education, Extension Methods, Communication and Diffusion Process, but also elaborately deals with Reforms. in Extension in India, Support to State Extension Programmes for Extension Reform, Rural Development & Community Development in India, Participatory Extension Planning and Management, Participatory Rural Appraisal, Participatory Learning and Action, Strategic Research & Extension Planning, Decentralised Planning & Micro-Level Planning and Gender Mainstreaming in agriculture for Rural Development.

I hope, the book would help the students, researchers, development practioners, policy makers and scholars aspirants for competitive examinations. to have thorough understanding of different concepts, methodologies and approaches for rural development through Agricultural Extension as a Special Purpose Vehicle.

I wish the publication a wider reach and all success.

Bhubaneswar, the 16th December, 2023

(Dr. Arabinda K. Pandhee)

Preface

With the shrinking of physical resources and increasing scarcity of land use products, such as food, fuel, feed, and fiber, along with increasing awareness regarding the risks associated with climate change, there have been great challenges for agricultural research, education, and extension. The textbooks on agricultural extension have not been updated in recent years, and they do not cover recent developments and reforms in Agricultural Extension. The Old Testament of the Bible as well as ancient writings about the Greek philosopher Socrates contains some astonishingly up-to-date extension wisdom; thus, 'old' does not always mean 'outdated'. However, the world has changed in many respects over the past thirty years, and this is also true with rural extension. In this sense, updating provides an opportunity to carefully check the validity of concepts, statements, and recommendations and their relevance for extension work today and tomorrow, thereby including recent developments and experiences as well.

Agricultural extension is one of the important drivers of farmers' development through the conscious use of communication and dissemination of useful technologies emanating from research stations. The role of extension field staff is to work as a carrier of innovations from the research to the farmers. The dissemination of innovations is impossible without the agricultural extension sector. Extension work requires a collaborative effort of both rural people and extension workers. The aim of the extension agency is to transmit reliable, authentic, and doable technologies, advice, and guidance for the socio-economic development of rural people. Further, extension is seen as a broad effort directed at rural development linking with other sectors in addition to the transfer of agricultural technologies.

Extension reforms to make it farmer-centric and participatory have been initiated in recent years focusing on strategic planning, implementation, monitoring, and evaluation. It is now realized that participatory extension management using various participatory tools will improve the decision-making capacity of the farmers. Village-level micro-planning and Strategic Extension and Research Planning (SREP) have become imperative in the extension reforms in India for which support is given by the Government of India to the States and UTs. Gender mainstreaming is another area that needs thrust in the extension management system. In addition to traditional extension methods, modern methods of extension have become cost-effective and sustainable. Since agriculture and rural development are intimately associated and mutually inclusive, the extension officers must be conversant on various Rural Development Programs.

Considering the above requirements of Agricultural Extension, this book has been written containing all aspects of Modern Extension Systems and Participatory Extension Approaches. The overall aim of the authors of this book is to present and disseminate the concepts and methodologies of agricultural extension and its relevance to rural development. The book attempts to explain basic theories and important relationships in the multitude of extension activities. The authors have tried to keep the text as simple as possible, making it easier to read by avoiding purely technical discussions. The purpose of this publication is to highlight the important role of agricultural and rural extension in the backdrop of Sustainable Development Goals and provide all information regarding Participatory Extension Management which will be useful for the students, practitioners, NGOs, policymakers, and extension mangers.

The Book contains 12 chapters. The first chapter deals with the Basics of Extension Education. Chapter II contains the Historical Evolution of Agriculture Extension in India. Chapters III to V deal with communication, extension methods, and adoption, diffusion, and innovations while Chapter VI deals with Reforms in Agricultural Extension in India. Chapter VII deals with Participatory Extension Management. Extension Programme Planning, Monitoring, and Evaluation are dealt in Chapter VIII. Rural Development and Community Development have been described in Chapter IX in detail. Participatory Rural Appraisal (PRA) and Participatory Learning and Action (PLA) have been explained in Chapter X while Chapter XI deals with Decentralized Planning and Micro-level Planning. The last Chapter (XII) deals with Gender Mainstreaming for Rural Development.

The authors have tried their best to bring in the readability of the book and incorporated recent developments and research findings quoting the authors in the book. The authors acknowledge the writers and researchers and the literature that have been referred to in writing this book. Suggestions are invited to include the omissions in the next edition of the Book. The book is expected to be useful to students, extension personnel, policymakers, and non-government agencies engaged in extension and rural development.

Bidyadhar Maharana
Chakradhar Satapathy

Contents

Abbreviations

AES	:	Agro-Ecological Situation
AGRISNET	:	Agricultural Resource Information System Network
AICRP	:	All India Coordinated Research Project
AMC	:	ATMA Management Committee
ATMA	:	Agricultural Technology Management Agency
AV	:	Audio Visual
BAP	:	Block Action Plan
BFAC	:	Block Farmer Advisory Committee
BTT	:	Block Technology Team
CBO	:	Community Based Organisation
CDP	:	Community Development Programme
CIG	:	Commodity Interest Group
CPA	:	Community Problem Analysis
CSC	:	Common Service Centre
CTTA	:	Communication for Technology Transfer in Agriculture
DAC	:	Department of Agriculture & Cooperation
DAP	:	District Action Plan
DASP	:	Diversified Agricultural Support Project
DRWA	:	Directorate of Research on Women in Agriculture
FAC	:	Farmer Advisory Committee
FAO	:	Food and Agriculture Organization of the United Nations
FFS	:	Farmers' Field Schools
FGD	:	Focus Group Discussion
FIAC	:	Farm Information and Advisory Centre
FIG	:	Farmer Interest Group

FPO	:	Farmer Producer Organisation
FSG	:	Food Security Group
GP	:	Gram Panchayat
HIV/AIDS	:	Human Immuno deficiency Virus / Acquired Immuno deficiency
HYVP	:	High Yielding Variety Programme
IAAP	:	Intensive Agricultural Area Programme
IADP	:	Intensive Agricultural District Programme
ICT	:	Information Communication Technology
IDWG	:	Inter Departmental Working Group
IPM	:	Integrated Pest Management
IRDP	:	Integrated Rural Development
ISD	:	Information System Development
ITD	:	Innovations in Technology Development
IVS	:	Indian Village Service
KCC	:	Kishan Call Centre
KVK	:	Krishi Vigyan Kendra
LDCs	:	Least developed countries
MLP	:	Micro Level Planning
NAEP	:	National Agricultural Extension Project
NAIP	:	National Agricultural Innovation Project
NASC	:	National Agricultural Science Centre
NATP	:	National Agricultural Technology Programme
NES	:	National Extension Service
NGO	:	Non Government Organisation
NMAET	:	National Mission on Agriculture Extension and Technology
ODI	:	Overseas Development Institute (U.K.)
PEA	:	Participatory Extension Approach
PGN	:	Practical Gender Needs
PLA	:	Participatory Learning and Action
PME	:	Participatory Monitoring and Evaluation

PNA	:	Participatory Need Assessment
PRA	:	Participatory Rural appraisal
RD	:	Rural Development
REF	:	Research Extension Farmer
RKMP	:	Rice Knowledge Management Portal
RKVY	:	Rashtriya Krishi Vikash Yojana
RPO	:	Rural Producer Organization(s)
RRA	:	Rapid Rural Appraisal
SAMETI	:	State Agricultural Management Extension and Training Institute
SAP	:	State Agricultural Plan
SAU	:	State Agricultural University
SEWP	:	State Extension Work Plan
SGN	:	State Gender Needs
SLSC	:	State Level Sanctioning Committee
SMAM	:	Sub-Mission on Agricultural Mechanisation
SMPP	:	Sub-Mission on Plant Protection and Plant Quarantine
SMSP	:	Sub-Mission on Seeds and Planting Material
SNC	:	State Nodal Cell
SREP	:	Strategic Research and Extension Planning
SWOT	:	Strength, Weakness, Opportunity and Threat
T&V	:	Training and Visit
TDU	:	Technology Dissemination Unit
TDMC	:	Technology Dissemination Management Committee
TOFA	:	Team of Farm Advisors
TOT	:	Transfer of Technology
UNESCO	:	United Nations Educational, Scientific and Cultural Organization
WFS	:	World Food Summit
ZP	:	Zilla Parishad
ZRS	:	Zonal Research Centre

1

Basics of Extension Education

Introduction

Extension is a form of education, and extension services provide information, advice, and education for rural development in many countries. It was used for sharing ideas, beliefs, knowledge, and *information* with rural people in an informal way. In the mid-**nineteenth** century, extension services were institutionalized, and public funding was spent for organized extension activity in rural areas.

Extension Education

The word 'extension' is derived from two Latin roots, namely 'ex' meaning 'out' and 'tensio' meaning 'stretching.' Hence, extension means stretching out Figure 1.1.

Extension does not have a universally recognized definition. According to Webster, education is a process of teaching to develop the knowledge, skills, and character of students. Sociologist Rodney Stark declares that "Education is the cheapest, most rapid and most reliable path to economic advancement under present conditions." Education is a lifelong process, and it is necessary to transfer experience and knowledge to the next generation for political and socio-economic development.

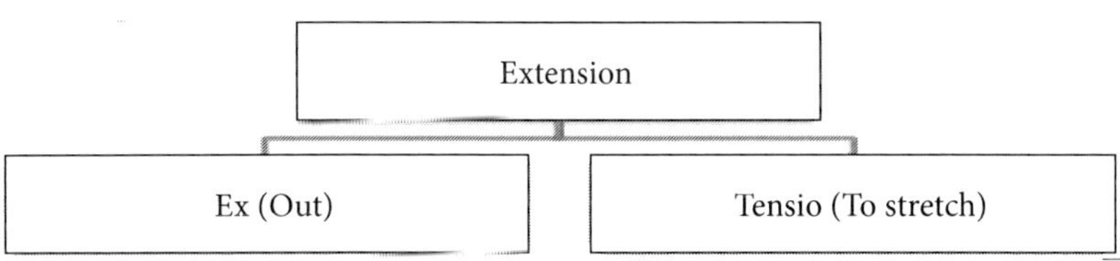

Figure 1.1: Extension education

University extension started at Cambridge University, London University, and Oxford University before the nineteenth century. The success of extension in Britain led to the initiation of extension education in the United States later. In India, university extension was introduced after the establishment of State Agricultural Universities in a similar pattern to Land Grant Colleges in the United States.

Extension as a form of education facilitates the learning or acquisition of knowledge, skills, values, and habits. As an education, it can be informal, formal or non-formal. It can also be categorized as child education, adult education, technical education, health education, social science and humanities, etc.

Formal Education: It is an institutionalized, chronologically graded, and well-structured system of education that starts from schooling to higher education. Formal education is usually given in structured classrooms in schools and colleges with a fixed syllabus, course curriculum, and certificates awarded after examination. In formal education, only qualified teachers impart education to the students.

Informal Education: It is the never-ending process by which an individual learns through daily experiences and exposure to the environment at home and at work. He learns from friends, radio, television, papers, books, etc. In informal education, the learner acquires attitude, values, skills, and knowledge from daily experience or education in their own environment. Informal education is participatory and voluntary. It is spontaneous without any pre-planning, and it does not use a timetable or and course curriculum.

Non-formal Education: It has been a well-organized, systematic educational activity that is carried out outside the formal educational system in order to provide certain selected types of learning to a selected group of individuals, including adults, young people and children. The subjects of non-formal education are agriculture and related fields, where training is given for the development of skills, attitudes, and knowledge. Home science activities such as health, nutrition, childcare, sanitation, fruit and vegetable preservation, etc. are covered under non-formal education.

Definition of Extension Education

Extension education is defined differently by many scientists and organizations according to their interests and objectives.

The goal of extension education is to teach rural residents how to live better by educating them on how to make improvements to their homes, farms, and community institutions (Leagans, 1961).

Ensminger (1962) defined extension as education, and its purpose is to change the attitude of the people with whom the work is done.

Extension is a school education that provides services to farmers or others engaged in farming directly or indirectly for the adoption of improved practices for production and marketing (National Commission of Agriculture, 1976).

Extension Education is a science that brings about desirable changes in the behavior of the persons concerned to improve their standard of living with their own efforts through education methods.

From the above definitions, it is apparent that extension education is an out-of-school non-formal education with the intention of providing services for agriculture and rural development. Extension process is a means of working with rural people through out-of school education with their current interests and needs, which are closely related to livelihood improvement and well-being, and fostering rural community welfare.

Objectives of Extension Education

The main aim of extension education is to bring about the all-round development of rural people, which includes educational, social, economic, and political developments. Agricultural extension is primarily concerned with the dissemination of useful and practical information relating to improved practices and inputs in the field of agriculture and allied subjects. It aims at the practical application of useful knowledge to agricultural farms, thereby ultimately improving all aspects of the lives of rural people. The extension workers need to assist the villagers in village planning for integrated development covering agriculture, livestock, handicrafts, village industries, and overall living standards.

Extension and Extension Education

Extension is a non-formal education aimed at bringing out a behavioral change in its clients for their increased knowledge, skills, and income. Roling (1982) explained that extension refers to the actual work done by professional extension agents at various levels of management, while extension education refers to the body of knowledge concerning that work practice.

Extension education is applicable to a number of areas, such as Agricultural Extension, Animal Husbandry Extension, Fishery Extension, Forestry Extension, Irrigation Extension, Farm Mechanization / Agricultural Engineering Extension, Home Science Extension, Medical Extension, etc.

Objectives of Extension Education

The basic objective of extension education is to improve the well-being of rural people and make them self-reliant. The specific objectives are:

- To improve the decision-making power of the rural people to solve their own problems,
- To assist the rural people in providing minimum health, recreational, and educational facilities for improving family living standards in the village;

- To increase the net income of farmers through more production through the adoption of better technologies, inputs, and agro-processing;
- Integrated development of rural areas by providing facilities for socio-economic development; and
- Developing rural leadership requires a feeling of self-dependence.

Scope of Extension Education

Extension has unlimited scope in creating awareness amongst rural people and changing their behavior, skills, and attitudes. In describing the scopes of Extension Education, Kelsey and Hearne (1967) recognized nine areas of program emphasis. These areas are: (i) efficiency in agricultural production; (ii) efficiency in marketing, distribution, and utilization; (iii) conservation, development, and use of natural resources; (iv) management on the farm and in the home; (v) family living, (vi) youth development; (vii) leadership development; (viii) community development and rural area development; and (ix) public affairs. The scope of extension is wide and diverse, which can lead to economic development, socio-cultural development, human development, and political development.

Philosophy of Extension Education

Extension is based on the philosophy of self-help, voluntary co-operation, democracy, and teaching through 'learning by doing' and 'seeing is believing.' It is participatory learning involving a two-way channel. Extension is an education process that is an economic necessity based on research and people's needs.

Principles of Extension Education

Extension work is based upon some working principles and knowledge that are followed by the extension workers.

- The extension education should depend on the felt needs and interests of people.
- Local people in groups should work with the community to adapt to local conditions.
- Extension workers should value the culture, tradition, values, and customs of local people and work in harmony with such customs.
- Extension should seek the cooperation and participation of the people in a democratic manner so that people voluntarily participate to create self-confidence.
- Extension should involve local leaders to help the extension workers to convince people to accept new ideas without any resistance.

- The extension program needs to be flexible to allow required changes as per local situations, and the extension workers should be conversant on various methods of extension so that the appropriate method could be selected as per the demand of the local situation and people.
- Extension should be a whole-family approach.
- The extension work should be based on helping people to help themselves and learn by doing.
- Extension education should aim at people's satisfaction in solving problems, fulfilling their needs, and acquiring new skills, or changing their attitude and behavior.
- As people's indigenous knowledge (IK) is essential for providing local solutions to most of their problems, the extension workers need to collect the (IK) and, if required, validate it in research stations for their replication.
- Extension needs to apply democratic principles for the adoption of new technologies as per discussion with them.
- Extension workers should always be neutral without showing any bias towards a group and should not take an interest in local politics.
- Extension should provide scope for constant evaluation of the effectiveness of the work in terms of change in knowledge, skills, attitude, and adoption behavior.

Process of Extension Education

There are five interrelated steps that are adopted for achieving success in extension education (Table 1.1).

Table 1.1: Five Interrelated Steps for the Process of Extension Education

Step	Process
First	Collection of data by appropriate survey and primary methods for situation analysis and identification of the gaps and problems.
Second	Setting goals and realistic objectives in consultation with the community for socio-economic development.
Third	Selection of appropriate methods to be adopted and deciding teaching methods and technologies based on research findings to address community problems.
Fourth	Evaluation of the results of success and collection of evidence.
Fifth	Reconsideration of the extension program based on the evaluation and identification of problems.

The cyclic process of extension education has been displayed in Figure 1.2.

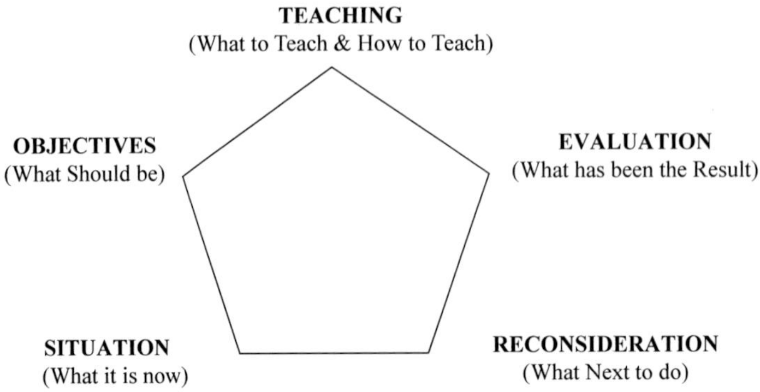

Figure 1.2: The Extension Education Process

Rationale for Extension Education

Extension education is considered necessary to identify the gaps and problems of rural people and to bring about a desirable change in the existing situation. The extension agency serves as a link between the people and research stations for the solution of problems faced by the community, as it may not be possible for direct contact between the research scientists and the people.

The following functions are usually discharged by the extension workers for rural development.

Change in knowledge, attitude, and skills of the people: Extension education is required to change the person's present knowledge by providing information about the latest developments, bringing a positive change in attitude toward the latest information given to them, and changing the skill to ensure better performance.

Change in understanding of technologies: The understanding level of people needs to be changed so that they will believe that the appropriate technology can solve their problems.

Change in goals of the people: Goal setting for higher productivity and profitability would be required in the agricultural sector and required technologies are required for this. This can be facilitated by the extension workers.

Change in action of the people: People should be advised by the extension workers to make plans about the implementation of new techniques and start action.

Change in confidence of the people: Confidence building among people is very important so that people can adopt new techniques by which there will be a definite increase in their net income. The extension work also needs to be confident about the efficiency of the technologies.

Developing leadership: Leadership development is required in villages to adopt new technology first and help spread across the village.

Development of rural groups and rural institutions: Group extension is required to ensure collective efforts. Through the proper development of the gram Panchayat, co-operative institutions, and other institutions, greater cooperation is achieved in rural development.

Barriers to Farmers

The common barriers to farmers, such as knowledge, motivation, resources, insight, and power, are overcome by extension.

Knowledge: The farmers require certain knowledge to solve their problems and achieve their goals. This knowledge is provided by extension workers.

Motivation: In the absence of proper motivation, some farmers do not change their behavior. For example, farmers often neglect critical practices in anticipation of extra expenditure and time. The extension workers need to change their behavior through proper motivation and demonstration.

Resources: Extension is required to facilitate the supply of key inputs like seeds, fertilizers, and credit, as input-supplying agencies are not engaged in extension. The extension workers can suggest appropriate quantities of inputs required and their sources of availability.

Skill: Adoption of technology would require adequate and appropriate skill. Hence, extension workers should transfer appropriate skills to the farmers. The extension worker has to demonstrate the manner of adoption of technologies.

Insight: Farmers may lack insight into how to avail themselves of the required resources, which the extension worker needs to pass on such information.

Power: Provision of information only may not change the power of farmers, and it is also not feasible with extension workers.

All the problems of farmers may not be solved by the extension workers, and scientific research may be required to solve many of the farmers' problems. However, extension workers must analyze the current situation so that they will be in a position to give timely information to farmers so as to solve their problems with appropriate knowledge, skills, and technology.

Extension and Government Policy

In public extension, it is difficult for the extension agency to work against government policy, even though some are against the interests of farmers. For example, the cultivation of *ganja* goes against government policy, while some farmers are interested in growing the crop in anticipation of huge profits in hilly areas.

Usually, the government's goals are: (i) increasing food production; (ii) stimulating economic growth; (iii) increasing the welfare of farm families and rural people; and (iv) promoting sustainable agriculture. Sometimes, these goals may conflict with each other. For example, increased food production is a short-term measure that cannot be sustainable in the long run, or it may reduce food prices and farm income. Extension is often used to achieve government goals that are not in the best interest of the farmers. This loses the confidence of farmers in extension workers, which makes extension ineffective. It is desirable that extension workers protect the interests of both the government and farmers.

Objectives of Extension Organization

There are many organizations, such as public, private, and Non-government organizations (NGOs) engaged in extension, and these organizations have their own objectives. For example, public extension would aim for higher production through the use of inputs and technologies, while NGO extension working in a small area might focus on indigenous technologies and organic farming. Public extension may work for increased production, while private extension may look at the profitability of agriculture. While choosing the objectives for extension program, the following considerations may be made:

Right to Decision making

Adoption of a new technology might increase return, and, at the same time, it may also increase risk. Then the farmer has to decide which is more important to him. Here, the extension might help the farmer to think about the consequences of the decision in a systematic way without persuading the farmer to make the same decision as per the advice of the extension.

Knowledge Required for Decision-making

In agriculture, the extension worker often has only part of the knowledge required for decision making, while the farmer and family have the other part. Therefore, the combined knowledge of the extension worker and the farm family has to be integrated to develop a more productive farming system for the family. The IK or indigenous technical knowledge of the farm family is very important to address their problems, which the extension worker should not ignore. There can be a blending of IK and exogenous knowledge.

Motivation to Implement the Decisions

The extension worker gives suggestions, and the farmer works on them. The extension worker should try to create a situation in which the farmer can make good decisions himself so that his decision-making skills will increase. In view of the variation in farm size, land quality, availability of resources, and goals in the same village, the farmers should make different decisions about different production technologies. Human resource development among farmers is very important.

Development Policy

The government wants to achieve development policy objectives through extension. But the extension agency should analyze the reasons for non-compliance with the objectives and then decide the action against knowledge, skills, motivation gaps, etc. A participatory relationship between the farmers and the extension worker requires a participatory style of leadership in the extension organization.

Changing Farms or Famers

Generally, technology is taught to extension workers on how to change the farm. However, their role is to change the farmer, who will adopt the new technologies on his farm. Changing farmers can be achieved through adult education and communication techniques. The extension organization receiving information from research institutes and policy level will have to be carried out to farmers by the extension organization through extension workers. Farmer-to-farmer transfer of such information should also take place, and the feedback of farmers should be taken back to research through extension.

Helping Farmers to Reach their Goals

The extension workers can use different ways, like giving timely advice, suggesting alternatives, facilitating analysis of different alternatives, helping them to make decisions as individuals or as groups, and exchanging information, among others.

Farmers' Organizations

Farmers' organizations play an important role in serving their collective interests and in agricultural development. Extension organizations can motivate the farmers to organize them effectively.

Educating Farmers

Farmers can be taught how to solve their own problems. They must be kept informed about new production technologies and markets. The extension agency has an important role in facilitating the process of self-directed learning.

Changing the Farmers' Situations

Extension agencies need to report the real situation of farmers to policymakers and research scientists. This might change the farmers' situations, or developing technologies might change the farmers' situations.

Agricultural Extension

Agricultural extension undertaken by government departments, universities, and other organizations has become an important instrument to advise farmers to make sound decisions on agricultural production systems. It has expanded its objectives with more focus on food and nutrition security and poverty alleviation through the involvement of many actors, such as NGOs. But public spending on extension services is shrinking day by day. The design of agricultural extension programs in developing countries has been the subject of heated debate for transformation. Many reforms on extension have been introduced in many countries for a cost-effective service in a public-private partnership mode.

Agricultural extension is necessary for the transfer of agricultural technologies to farmers to bring about a desirable change in their knowledge, attitude, behavior, and decision-making capacity. Agricultural extension covers the supply of material technology (seeds, fertilizers, etc.), the transfer of better technology for increasing crop and livestock productivity, establishing a linkage between farmers-extension-research, and encouraging adaptation and diffusion of innovations. Agricultural extension is a non-formal adult education that places emphasis on the communication of new technologies through appropriate extension methods. Since agriculture is regarded as the most effective means of poverty alleviation, agricultural extension assumes importance.

Extension Ethics

Both the ability of the extension workers in communication technology and the use of social science in extension are required to influence the farmers. The behavior of an extension worker is influenced by organization in which s/he works the farmers, professionalism, and society (Figure 1.3).

The extension worker needs to be loyal to the farmers, society, and their own profession and organization, which often becomes a stupendous task. The extension workers should help the farmers make decisions that are helpful to them. They should promote the autonomy of farmers in deciding how to develop their farming. After all, they should be honest. The extension worker must consider the above dimensions when deciding their relationship with farmers. The important qualities of a good extension worker are indicated below.

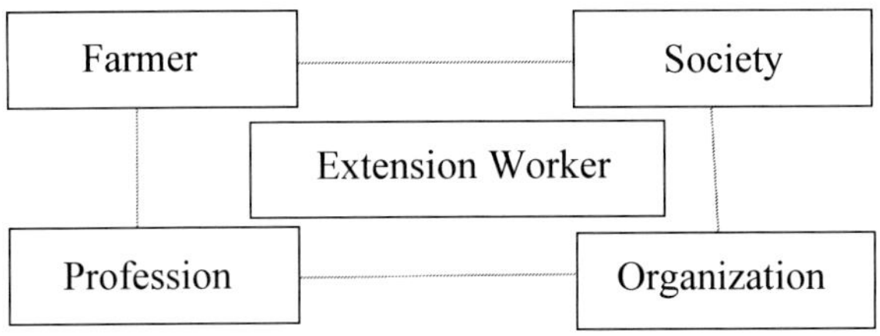

Figure 1.3: Extension worker's square of loyalties.

- He should be acquainted with the real problems of farmers and should have the knowledge to solve these problems.

He must be honest, sincere, hard-working, and empathetic toward the farmers.

- He should have a positive mind and vision.
- He should respect the farmers and be friendly with them.
- He should respect the traditions, culture, and social norms of the farmers.
- He should be patient and avoid agreeing with the farmers.
- He should search for new ideas and techniques by searching the literature or discussing them with research scientists.

References

Ensminger, Douglas. (1962): A Guide to Community Development,. New Delhi: Ministry of CD and Coop. Govt. of India.

Kelsey, L.D. and Hearne, C.C. (1967). Cooperative Extension Work, Cornell University, Press, Ithaca, New York.

Leagans J.P. (1961), Characteristics of Teaching and Learning in Extension Education, Extension Education in Community Development. Directorate of Extension, Govt. of India. New Delhi.

Röling N. 1986. Extension and the development of human resources: The other tradition in extension education. In: Gwyn EJ (ed), Investing in rural extension: Strategies and goals. Elsevier, London, UK.

Suggested Reading

A.W. Van Den Ben and H.S. Hawkins 1994. Agricultural Extension. Long man Scientific & Technical Publication. U.S.A.

Benor, D. and Harrison, J.Q. 1977. Agricultural Extension- Training and Visit System, World Bank, Washington.

Dahama, O.P. 1973. Extension and Rural Welfare, Ram Prasad and Sons, Agra.

Dahama, O.P. and Bhatnagar O.P. 1987. Education and Communication for Development. Oxford and IBH Publishing Co. Pvt. Ltd., New Delhi.

Directorate of Extension 1961. Extension Education in Community Development, Ministry of Food and Agriculture, Govt. of India, New Delhi.

Ganesh, R., Mohammad lqbal, 1. and Anandaraja, N. 2003. Reaching the Unreached - Basics of Extension Education. Associated Publishing Company, New Delhi.

Jalihal, K.A., & and Veerabhadraiah, V. 2007. Fundamentals of Extension Education and Management in Extension. Concept Publishing Company.

Mondal, S. 2019. Fundamentals of Agricultural Extension Education. Kalyani publishers, New Delhi.

Supe, S.V. 1983. An Introduction to Extension Education, Oxford and IBH Co., New Delhi.

Swanson, Burton, E. Ed. 1984. Agricultural Extension. A Reference Manual, Rome, Food and Agricultural Organization (FAO), of the United Nations, 2nd Edition.

2

Historical Evolution of Extension in India

India faced a series of famines from 1857 to 1901, for which the government was forced to appoint some commissions. The Famine Commission of 1880 recommended the establishment of a separate Department of Agriculture at both the central and provincial levels. Based on such recommendations, the Agricultural Departments were established in 1981–1982, and some acts like the Land Improvement Act and Agricultural Loans Act, 1988 and Co-operative Act, 1904 were passed and development departments were established. As a protection measure, some irrigation projects were created. The Famine Commission (1901) also recommended for establishment of Agricultural Research Stations and Colleges. The Imperial Agricultural Research Institute along with an Agricultural College was established at Pusa Institute. As per the Government India Act of 1919, all departments connected with rural development to the major provinces and agricultural development became a State subject. Rural Development works were made provincial subjects from 1935. Various diversified efforts for rural reconstruction were made for the development of extension in India during the Pre-Independence and Post-Independence periods. These extension programs were launched in four stages as indicated below.

Stage I: Pre-Independence Extension

Stage II: Post-Independence Extension (1947–1953)

Stage III: Community Development and National Extension service (1953–1960)

Stage IV: Intensive Agricultural Development (1960 onwards)

Pre-independence Extension in India

Between 1903 and 1946, eight important area-specific programs were implemented for rural development while the Grow More Food Campaign was implemented in the entire country to increase food grains production. The description of these programs/projects is given below in brief.

Ideal Village Program (1903)

The program was initiated by Sir Daniel Hamilton in Sundaram Village near Madras. The objective was rural development on the basis of cooperation for which a Cooperative Deposit Institute was established for savings, health, literacy, and small industries. Specialized training was also given to unemployed young farmers. The program made young farmers self-dependent for 16 years.

Work at Shriniketan (1908) and Rural Reconstruction Institute, Shriniketan (1921)

Shri Rabindra Nath Tagore started Rural Development work in 1908 in some villages of Kaligram Pragana of his Zamindari by creating some youth organizations with the help of K. Elmhirst. A Rural Reconstruction Institute was established in 1921 at Shantiniketan and it was called Shriniketan. Development of agriculture, cooperatives, industries, and education through village organizations was taken up in a group of eight villages focusing on the slogan of 'Education for all'.

The objectives of the Institute were to create interest in rural welfare work among people, study rural problems and take action, resource development of villages, and improve sanitation. A spirit of self-help for mutual help was created and leadership was developed along with organizing village scouts and establishing training centers and a demonstration center. In the demonstration center, improved practices, dairy, and poultry farming were taken up. The students and workers from the institute trained weavers, organized their cooperatives, and provided facilities for training in tanning, pottery, embroidery, tailoring, etc. The center was a good learning platform.

Gandhi's Sewa Gram Project (1920): Mahatma Gandhi believed that India lives in villages and he emphasized that the "Salvation of India lies in Cottages". He emphasized on self-sufficiency in Indian villages. With this ideology, he started the welfare project to create patriotism among them in in Ashram at Vardha. The project aimed at wearing Khadi cloths, implementing health, sanitation, women's welfare, economic help and upliftment of backward classes, and primary education. Several associations like the Village Industry Association, All India Spinners affiliation, Hindustani Education Association, Kasturba Gandhi Association, etc. were set up by Gandhiji to make the program successful. The program was not followed completely by the majority as it could not compete with the industry. However, the program was a people's movement by the people and for the people.

Gurgaon Project (1920): A rural uplift project was first started by Mr. F. L. Brayne, Deputy 'Commissioner in the Gurgaon district of the Punjab. A village guide was posted in each village for the dissemination of information and knowledge on improved seeds, implements, improved methods of cultivation, prevention of cow dung use as fuel, preparation of manure pits, and reduction of expenditure on social and religious functions. The program got momentum after 1933 when Mr Brayne was appointed as Commissioner of Rural Reconstruction in Punjab.

Marthandam Project (1928): Young Men's Christian Association (YMCA) set up this project in Travancore for the development of spirit, mind, and body for spiritual, mental, physical, and socio-economic development. This was a pioneer work of Dr. Spencer Hatch, an American Agricultural expert. Nearly 100 villages were covered by YMCA centers in villages under the supervision of an extension secretary. Awards were given for bulls and goats, model bee hives, demonstration plots for improving grain and vegetable seeds, poultry farms and weaving sheds, etc. In these centers, several equipments were kept to facilitate honey bee cultivation, poultry, etc. After the tremendous success of the Egg Selling Club, the Egg Selling Society was a Governing Body by 1939. Many health programs, exhibitions and lectures were arranged by the centres.

Baroda Village Reconstruction Project (1932): Started by Shri B.T. Krishnamachari in Baroda, the project aimed at the improvement of rural people, development of agriculture, education, and industrialization. Programs such as gardening, poultry farming, beekeeping, spinning and weaving, and education were implemented by village guides and teachers.

Grow More Food (1942): In order to meet the food grain deficit, the Grow More Food campaign was launched in 1942 throughout the country after the Second World War till independence. The government distributed key inputs like seeds and fertilizers free of cost to the farmers to increase production through better cultivation. The program responded with increased food grain production duc to the use of improved seeds and fertilizers. Although the program was launched in the entire country only limited farmers were provided support. The objectives of the program were re-defined after independence for self-sufficiency in food grains by 1952.

Indian Village Service (IVS): Mr. Arther T. Mosher of New York, and Shri B. N. Gupta established the Indian Village Service in 1945. The objective of the program was to assist the villagers in developing individuals, volunteer leaders, and local agencies, and enabling them to be effective in helping themselves and others. The program started in Agasoli village of Aligarh district in UP, Extension techniques/methods like personal contacts, informal group discussions, use of volunteers, demonstrations, production and use of visual aids, exhibitions, field trips, dramas, books, periodicals, etc. were used for communication. The program was implemented through contributions and donations. The organization was affiliated with the Presbyterian Church in the USA, and the implementation was made in 15 villages. There was limited success as the program was guided by foreigners and run by Christians.

Firka Vikas Yojana (1946): As the biggest project in the pre-independence era, it was implemented by Govt. of Madras. The objective was all-around village development for which specific schemes for the drinking water, cooperation, animal husbandry, farming and irrigation, khadi, and cottage industries were implemented. This was a successful program at the Firka (village) level in 84 Firkas till 1950.

Post-independence Programs

During the post-independence era, many integrated programs/projects were launched on an area-development basis or farmer-development basis. The real extension projects started with the introduction of the National Extension Service in 1953 which was improved subsequently by implementing the Training & Visit System (T & V System) in 1977 under external aids. Some projects were also launched for the training and employment of rural youth. The Programs/Projects launched after Independence are described below in brief.

Etawah Pilot Project (1948)

The project was started in the Etawah district of UP in 1948 under the guidance of Lt. Col Albert Mayer as a new experiment in rural planning and development. A team of four specialists was formed including a town and village planner, an agricultural extension specialist, an agricultural engineer, and a rural industry specialist for the integrated project. They were advising the people on the development of cottage industries, agriculture, stocking of seeds & fertilizers, etc. as inputs of agriculture, improvement of sterile land, prevention of soil erosion, and reconstruction of tube wells, health, education, and cooperation. The broad objectives were developing self-reliance, increasing agricultural production, and improving health and literacy at the village level. The project had an organized administrative mechanism at the state, district, and village levels. The project was an extension project which emphasized on local-level planning, teamwork, and communication of technologies to respond to local needs. The project started with 64 villages which were expanded to 400 villages by 1956.

Nilokheri Project (1948)

The project was started as a Refugee Rehabilitation Project in 1943 by Sri S.K. Dey but became functional from 1948. The main objectives of this project were: to rehabilitate 7000 displaced persons from West Pakistan; to attain self-sufficiency by utilizing the labor and talents of the refugees and to meet the basic needs of the population. With the establishment of vocational training-cum-production centers, training was imparted to create employment opportunities for the refugees. A new town named Mazdoor Manzil was established to accommodate 5000 people with facilities like hospitals, schools, vocational and technical institutions for training, veterinary and agricultural extension services, and recreational facilities.

Community Development and National Extension Service

Community Development Program (CDP): The Community Development Program (CDP) was launched on October 2, 1952 in phases under different plan periods with the assistance of three famous American specialists, namely Douglas Ensiminger, Chester Bowls, world-famous sociologist Karl Tailor. This was a multipurpose extension program that included on-farm and non-farm programs.

CDP is a balanced program for stimulating the local potential for growth in every direction. The broad objective of the program is to prepare a plan for the village, block, and district for increasing agricultural production, improving existing village industries, providing basic health and education facilities, providing recreation facilities, improving housing and living conditions of people, etc. and to implement the plan at Panchayat level. Focus was given to increased agricultural production by scientific knowledge, development of small irrigation projects, integrated community development, construction of roads and introduction of cooperative organizations, adult education, primary health service, cottage industry, provision of recreation facilities, and specific program for women and village youth. The project emphasized on individual development, family development, and rural leadership. Community problems were tackled by the community in a participatory manner. It laid the foundation for the extension program of agriculture in India.

National Extension Service (1953): National Extension Service (NES) program was launched in 1953 as a permanent program with a small budget. Due to the limitation of resources, Government decided to launch the National Extension Service program alongside community development projects. CDP represented on average 100 villages, with a population of 60,000–70,000 persons spread over an area of 150–170 square miles. However, the NES blocks were not developed with the same intensity. Out of the areas developed as National Extension Service blocks, the selection was made periodically for intensive development work under the community development program, and the blocks that were selected were named C.D. blocks. The NES block was headed by a Block Development Officer, who was assisted by a number of Extension Officers and about 10 multi-purpose village-level workers. NES intended socio-economic transformation of the rural people with limited funds and personnel. Both CDP and NES were complementary and interwoven and ran concurrently. After three years, the blocks covered under NES were converted into CDP blocks. The Extension officers were responsible for the implementation of technical programs in their respective subjects. NES worked in three stages before democratic decentralization in 1953–1958. These stages were pre-intensive, intensive and post-intensive. In the pre-intensive stage of three years, the people were made aware to create interest in development works. In the intensive stage, infrastructures were created in every block and specialists were posted. All pending works were completed in the post-intensive stage.

Intensive Agricultural Development (1960 onwards)

In order to eradicate poverty through economic upliftment, various agricultural development programs were launched in the 1960s. To start with Intensive Agricultural District Program (IADP), Intensive Agricultural Area Program (IAAP), High Yield Variety Program (HYVP), and Multiple Cropping Program (MCP) were implemented in selected districts of the country. Subsequently, the National Agricultural Extension Project (NAEP) and National Agricultural Innovation Project (NAIP) were implemented for agriculture development.

Intensive Agricultural District Program (IADP)

IADP was introduced in seven districts in 1960–1961 and subsequently extended to other nine districts in 1963–1964. In this program progressive fields were selected and the reliability of scientific agriculture technology was examined. The fertilizers, crop protection, improved seeds, adoption of improved farming practices, agriculture, and work of marketing were given to private sectors.

For intensification of crop production in selected areas, IAAP was launched in 114 districts in 1964-1965. Thereafter, the High Yielding Variety Program (HYVP) was launched in 1966-1967 with the introduction of dwarf wheat and rice varieties and other selected crops. The program was named "Green Revolution" after the Scientist William Gad of the USA. It was adopted by the government of India with the contribution of Dr M.S. Swaminathan, C. Subramaniam, and B. Sivaraman under the dynamic leadership of the then Prime Minister.

National Demonstration (1965): The National Demonstration Project (NDP) was launched by the Indian Council of Agricultural Research (ICAR) in 1965 with an aim to conduct crop demonstration fields by the Subject Matter Specialists on new technologies. National Demonstrations were conducted in four stages, and in the third stage, in 1969, these demonstrations were introduced in all districts intensively.

Command Area Development Program (CADA): With a view to increasing water use efficiency in irrigated command areas, the Command Area Development Agency was created in 1966-67. The purpose of CADA was to ensure coordination among irrigation, soil conservation, agricultural extension, and co-operation & credit agencies. The Agency took various steps like reduction of wastage of canal water, the introduction of better crop planning, control of soil erosion, construction of field channels, land levelling, and conjunctive use of surface and groundwater to increase the water use efficiency and reducing loss of canal water. Water use efficiency and crop productivity increased in command areas through this program.

Farmers Development Projects (1967)

After area development programs, farmers' development programs were launched focussing on small and marginal farmers including landless farmers. Small Farmers' Development Agency (SFDA) and Marginal Farmers' and Agricultural Labourer Development Agency (MFALA) were launched in 1967. The total expense of this project was borne by both central and state governments in a sharing mode. The objective of the Agencies was to provide institutional loans and subsidies for small irrigation, new farm machinery, and field projects. Identifying small farmers and investigating their problems, formulating economic programs for them, promoting rural industries, evolving institutional, financial, and administrative arrangements, etc. were the main functions of SFDA.

MFALA provided support for small industries for employment generation. Those farmers and laborers, who had smallholdings of less than one-hectare land or they got 50% of their income or more through agricultural labor, were benefited from this program. The district magistrate was the Chairman of this registered council and the Planning officer was the Secretary (designated as Project Director). These programs were constrained by limited coverage and the absence of local-level planning and monitoring.

Minor irrigation, soil conservation, land development, land shaping, animal husbandry, crop husbandry, the introduction of high-yielding varieties, and improved agricultural implements were projects supported for Marginal Farmers while providing adequate opportunities for offseason employment, poultry-keeping, piggery, sheep and goat-rearing, etc. were supported for agricultural laborers. A subsidy of 33.3% was given to the MF and 50% to agricultural laborers.

Training and Visit (T & V) System of Agricultural Extension (1977)

As a new extension reform, the Training and Visit (T&V) System of Extension was piloted in Rajasthan in 1974 following the concept of Dr Daniel Benor (World Bank Consultant) with funding support of WB. After initial success in Rajasthan, the system was expanded to other 16 states. The T&V System had a hierarchical organization with one Village Extension Worker for about 600-1000 farmers and eight VEWs are supervised by one Agricultural Extension Officer (AEO). Eight AEOs are supervised by one DAO/ADAO with the assistance of two Subject Matter Specialists (SMS). The DAOs/ADAOs of a district were supervised by the Principal Agriculture Officer (DDA) with the assistance of 3-5 SMSs. The VEO had the assistance of 80-100 contact farmers to reach 600-1000 farmers in his circle. The task of input and credit supply was entrusted to separate agencies. A single line of command and professionalism to transfer crop production technologies with research-extension linkage was responsible for the success of the system. The impact of the T&V system was readily apparent over a wide range of agroecological conditions in farmers' fields, whether irrigated or rain-fed. Regular training of the VEOs by the SMSs and regular visits to the farmer groups by the VEOs along with strict monitoring were the keys to the success of the system, which was reflected in the increased productivity of major crops.

The T & V system also faced some difficulties. It was a top-down system and the Village Agricultural Workers (VAWs) were asked to transfer technology which they learnt from the specialists. The system lacked in flexibility to adjust to local situations and was not participatory. The input dealers could not supply the inputs that were demanded by farmers after getting messages from the VAWs. After the withdrawal of WB support, the system faced financial difficulties in meeting the establishment expenditure. The unpaid contact farmers did not take an interest in serving as a link between farm families and VAWs.

National Agriculture Research Project (1979)

National Agricultural Research Project (NARP) is a research project that was started by ICAR with WB support to strengthen the research system of State Agricultural Universities. In this project, the agriculture-related problems associated with different climates was studied at the regional level to work out solutions through research interventions. The Zonal Research Stations (ZRS) were strengthened by the project to enhance their research capabilities. The researches were conducted for food grain and oilseed production under un-irrigated situations, on animal husbandry, water and soil conservation, and various agronomic methods. Appropriate land use pattern was followed for the use of natural resources and ecological potential in the region.

Transfer of Technology Program (1985)

With a view to transferring technologies emanating from research stations to the farmers' fields Lab to Land Program (LLP) and Operational Research Project (ORP) were launched. In 1985, the Indian Agriculture Research Institute took a challenging responsibility to transfer the latest technologies to the farmers' fields. The farmers have been freely provided with improved seeds, fertilizers, and agricultural chemicals, along with the technical knowledge of crop production by technical workers. Krishi Vigyan Kendras or Farm Science Centres were established at the district level by SAUs, ICAR Institutes, and reputed Volunteer Organizations to provide vocational training to farmers and extension workers and conduct front-line demonstrations. Currently, the KVKSs are conducting field experiments, farmers training, training to extension staff, and production and distribution of important seeds and planting materials.

National Agricultural Technology Project (1998)

After the closure of the T&V system of Extension, the National Agricultural Technology Project was introduced with WB support in 1998 as a pilot in selected 28 districts of seven states. The government of India implemented the project with the assistance of the National Institute of Agricultural Extension Management (MANAGE), Hyderabad in the selected states for a period of five years initially. MANAGE had earlier experience supporting the Diversified Agricultural Support Project in UP which was utilized in NATP. The three goals of NATP are decentralized decision-making at the district level by creating an Agricultural Technology Management Agency (ATMA) as a registered society, increasing farmer input into program planning and increase accountability of stakeholders, and increasing program coordination and integration.

The diverse activities are planned under three major components: (i) Agro-eco System Research; (ii) Innovations in Technology Dissemination (ITD); and (iii) Organization and Management Systems. The ITD proposals given by the Department of Agriculture & Cooperation and ICAR are processed through the Technical Dissemination Unit (TDU) and approved by the Technology Dissemination Management Committee (TDMC).

Selected State Agriculture Management and Extension Training Institutes (SAMETIs) were strengthened to train farmers and extension workers. Under the O&M component, the activities supported were (a) organization and management (O&M) reform in the ICAR; (b) strengthening of the ICAR headquarters, including the National Agricultural Science Centre (NASC); (c) Directorate of Information and Publications of Agriculture (DIPA); (d) institutionalization of priority-setting mechanism, monitoring, and evaluation (PM & E); and (e) Information System Development (ISD), covering Agriculture Research Information Service (ARIS) and Library Information, and Networking.

After the project became successful in pilot districts the project was later expanded throughout the country in 2005.

Agricultural Technology Information Centre (ATIC): The Agricultural Technology Information Centre (ATIC) was launched in 1999 under NATP to provide information on agriculture technologies in addition to providing other inputs like seeds, plant materials, etc. including advisory services through a single-window system. ATIC is a "Single window" support system linking the various units of a research institution with intermediary users and end-users (farmers) in decision-making and problem-solving exercises.

National Agricultural Innovation Project (NAIP)

ICAR has been implementing the project since 2006 with credit support of WB with the broad objective of facilitating the sustainable transformation of Indian agriculture for rural poverty alleviation and income generation through agricultural innovations. This was a collaborative project of public research organizations, farmers' groups, NGOs, the private sector, the civil societies, and other stakeholders. The NAIP attempted a sustainable transformation of the Indian agricultural sector to more of a market orientation to relieve poverty and improve income. Strengthening the National Agricultural Research System (NARS), research on production-to-consumption systems, conducting research on rural livelihoods security, and supporting research in frontier areas of agriculture are the four components of the project.

National Horticultural Mission (NHM)

Horticulture covers fruits, vegetables, flowers, spices, roots and tubers, medicinal and aromatic crops, cashew and cocoa, and oil palm including mushroom and apiary. In order to ensure holistic growth in the horticultural sector by increasing area and production through incentives and modern technologies, the National Horticultural Mission (NHM) was launched in 2005-2006. All expenditures are borne by the Government of India and State

Governments in a sharing mode. The Mission focuses on planting materials, production technology, post-harvest technology, and processing & value addition. Incentives are given to farmers for production and productivity improvement, production and distribution of planting materials, protected cultivation, tissue culture, vegetable seed production, seed infrastructure, creation of new gardens, creation of water sources, promotion of integrated pest management/integrated nutrient management, organic farming, etc. Currently, NHM is a sub-scheme of the Mission for Integrated Development of Horticulture (MIDH) which is implemented by the State Horticultural Mission.

Rashtriya Krishi Vikas Yojana (RKVY)

Rashtriya Krishi Vikas Yojana (RKVY) was launched in 2007–2008 by the Government of India to ensure at least 4% agriculture growth by preparing a comprehensive agricultural development plan at the state level. The scheme was implemented as a State Plan Scheme till the end of the financial year 2013-2014 and is being implemented as a Central Sector Scheme (CSS) and State Plan scheme thereafter. The funding pattern of the scheme from 2015 to 2016 onwards has been changed to a ratio of 60:40 between the Centre and States (90:10 for North-Eastern and Himalayan States).

RKVY has re-structured as RKVY Cafeteria Scheme from 2022-23 onwards merging some schemes of the Department of Agriculture & Farmers' Welfare which includes Soil Health & Fertility, Rainfed Area Development, Paramparagat Krishi Vikas Yojana (PKVY), Per Drop More Crop, Agriculture Mechanization (including Promotion of Agricultural Mechanization and Management of Crop Residue (CRM), Village Haats & GRAAMS and Crop Diversification Program. There are three components, an Annual Action Plan (AAP), a Detailed Project Report (DPR), and Administration, Monitoring, and Evaluation including a start-up in the cafeteria of RKVY. Out of the total allocation of RKVY, a minimum of 70% will be allocated for the Annual Action Plan (AAP) component including Administration, Monitoring, and Evaluation purposes, and a maximum of 30% will be earmarked for Detailed Project Report (DPR) component including Administrative charges and Star-up Components.

Agri-clinic and Agribusiness Centres Scheme (ACABC)

The scheme was started from 2000-01 to transfer the latest research knowledge on agriculture directly to the farmers. Agri-business Centres are commercial units of agri-ventures set up by skilled agriculture professionals to give advice to the farmers. NABARD is performing as a subsidy channelizing business enterprise. The objectives of the ACABC are to supplement extension and other services to farmers either on a payment basis or free of cost as per the business model of agri-preneur, to guide agricultural development, and to create gainful self-employment opportunities for unemployed agricultural graduates, agricultural degree holders, intermediate in agriculture and science graduates. Agri-Clinics offers expert advice to farmers on various technologies to enhance the productivity of crops/animals and increase the incomes of farmers. Commercial ventures may include maintenance and custom hiring of farm equipment, sale of inputs and other services in agriculture and allied sectors, post-harvest management, and marketplace linkages for income generation and entrepreneurship improvement. The scheme covers complete financial aid for training and handholding support, provision of loans, and credit-linked and back-ended subsidies.

Graduates in agriculture and allied subjects from State Agriculture Universities (SAUs)/Central Agricultural Universities/Universities recognized by ICAR/UGC are the beneficiaries. Diploma holders, Biological Science graduates, and Intermediates in agricultural subjects can also be the beneficiaries after approval of the Department concerned.

Table 2.1: Milestone of Agricultural Extension in India After Independence

1950s	Community Development Program (1952) and National Extension Service (1953) were the first planned attempts to educate farmers about improved methods of farming.
1960-66	The area-based special programs like Intensive Agricultural District Program (IADP), Intensive Agricultural Area Program (IAAP), and the High Yielding Varieties Program (HYVP) were launched to increase agricultural productivity.
1967	Farmers Training Centres (FTC) were created to educate farmers about high-yielding varieties and train them in improved methods of farming to augment the above programs.
1967	Small Farmers Development Agency (SFDA) and Marginal Farmer and Agricultural Labourers Development Agency (MFALA) launched projects to provide institutional loans and subsidies for small irrigation, new farm machinery, and field projects to small farmers and support for the small industry for providing more employment to MF and Agricultural laborers.
1974	World Bank's Training & Visiting (T&V) program was introduced as a pilot program in Rajasthan.
1977	World Bank's Training & Visiting (T&V) program was scaled up to 28 districts of seven states and then it was extended to many other states and districts.

1998	The National Agricultural Technology Project (NATP) was launched in 1998 under the financial support of WB as a pilot project in 28 districts of seven states for five years under the guidance of MANAGE after which it was extended to other districts and states. The district-level Agricultural Technology Management Agency (ATMA) was established under the ITD component to implement the project with a cafeteria of activities at the district level.
2014-2015	In order to make the extension system farmer-driven and accountable by restructuring and strengthening existing agriculture extension programs the National Mission on Agricultural Extension & Technology (NMAET) was launched in 2014-2015. It consists of four Sub Missions, such as Sub Mission on Agricultural Extension (SMAE), Sub-Mission on Seed and Planting Material (SMSP), Sub Mission on Agricultural Mechanization (SMAM), and Sub Mission on Plant Protection and Plant Quarantine (SMPP)

In addition to public extension services, various non-public players are also involved in agricultural extension. Such extension is undertaken by input dealers, agri-business companies, NGOs, Media and ICT, etc.

Suggested Reading

Aslam, M., and R.P. Singh. 1995. Evolution of Panchayati Raj and the Constitution (73rd Amendment) Act, 1992. New Delhi: IGNOU.

A.W., Van Den Ben, and H.S., Hawkins. 1994. Agricultural Extension. Longman Scientific & Technical Publication. U.S.A.

Babu, S.C., Joshi, P., Glendenning, C.J., Kwadwo, A.O., and Rasheed, S.V. 2013. "The State of Agricultural Extension Reforms in India: Strategic Priorities and Policy Options." Agricultural Economics Research Review 26: (347-2016–17086): 159–172.

Benor, D., and Harrison, J.Q. 1977. Agricultural Extension - Training and Visit System. Washington: World Bank.

Choubay, B.K. 1979. A Handbook of Extension Education, Jyoti Prakashan, Allahabad.

Dahama, O.P. 1973. Extension and Rural Welfare, Ram Prasad and Sons, Agra.

Dewan, M.L. 1982. Agriculture and Rural Development in India. New Delhi: Concept Publishing Company, New Delhi.

Directorate of Extension 1961. Extension Education in Community Development, Ministry of Food and Agriculture, Govt. of India, New Delhi.

Gulati A, Sharma P, Samantara A and Terway P. 2018. Agriculture extension system in India: Review of current status, trends and the way forward. Indian Council for Research on International Economic Relation.

Nandi, R, and Nceduraman, S. 2019. "Agriculture Extension System in India: A Meta-Analysis." Research Journal of Agricultural Science 10 (3): 473–479.

Singh, S.N. et al. 1991. Transfer of Technology to Small Farmers. New Delhi: Concept Publishing Company.

Swanson, B.E. 2008. Global Review of Good Agricultural Extension and Advisory Practices. Rome: Food and Agriculture Organization of the United Nations.

3

Extension and Communication

Introduction

Communication is a process for exchanging ideas, facts, feelings, or impressions between two or more individuals for a common understanding. Communication is required for transferring technology or obtaining farmer feedback to adopt technology in the field. Good communication is the essence of good teaching. Communication is a whole learning process in which the teacher sends a message through a channel to the receiver as a means of teaching. In traditional extension, there was direct contact between the extension worker and farmer to communicate research innovations and to obtain farmer feedback. In this system, only a few farmers had direct contact with the extension worker while the majority of farmers were contacted less frequently. Daniel Benor has tried to improve traditional extension through a "training and visit" (T & V) system designed to emphasize the organizational and training components of extension and improve communication with the farmers. In considering extension as a vehicle for communicating with large numbers of farmers without any discretion, the communication process and its role in extension need to be followed by the extension workers.

What is Communication?

Communication is a process of transferring a message from a source (communicator) through a channel to the receiver (communiqué). It is defined by different scientists in a different manner keeping the intent in view. However, it is accepted that communication is a process where a message is transferred from the source (communicator) to the receiver (communiqué). Some definitions of communication are indicated below.

- J. Paul Leagan (1961) defines communication as a process where two or more individuals exchange ideas, facts, and impressions for mutual gain and common understanding of the content and use of a message.
- "Communication is a purposeful process, which involves sources, messages, channels, and receivers." – Andersch *et al.* (1969).

- Van den Ban and Hawkins (1988) defined communication as the process of sending and receiving messages through channels that establish a common meaning between a source and receiver.

From the above definitions, it is seen that communication is a process of sharing information, ideas, attitudes, and likes with others. Communication involves six elements such as communicator, message, channel, treatment, audience, and response of the audience.

For communicating the ideas, a message is sent to the receiver from the source through a channel. Before understanding the proper meaning of the message the receiver has to decode that concept in his mind and he has to decide in his mind for this concept, whether this concept should be used or not. As the process involves the sender and receiver, the message will be successful when there is a common understanding between both. In the communication process, the receiver sends feedback to the sender. The process of communication is represented diagrammatically in Figure 3.1.

The communication process can be verbal or non-verbal depending upon the channel used to send the message. Verbal communication may be written or oral whereas non-verbal communication makes use of body language, facial expression, diagrams, sign-boards, paintings, or pictures. Body postures, hand gestures, and overall body movement are non-verbal communications. In verbal communication, the information is given directly either face to face or through telephone, internet, or SMS. Usually person spends 10–11 hours every day in verbal communication, which includes speaking, listening, reading, or writing. Conversations, speeches, letters, newspapers, magazines, audio, and video are included in verbal communication.

The purpose of communication is sharing information which involves control and motivation. Communication acts to control behavior of persons in several ways. Organizations communicate guidelines to the employees for necessary follow-up. Communication can foster motivation by clarifying doubts about the message and it sends information to individuals or groups for decision-making and evaluating alternative choices. Good communication is required both for research and extension in agriculture.

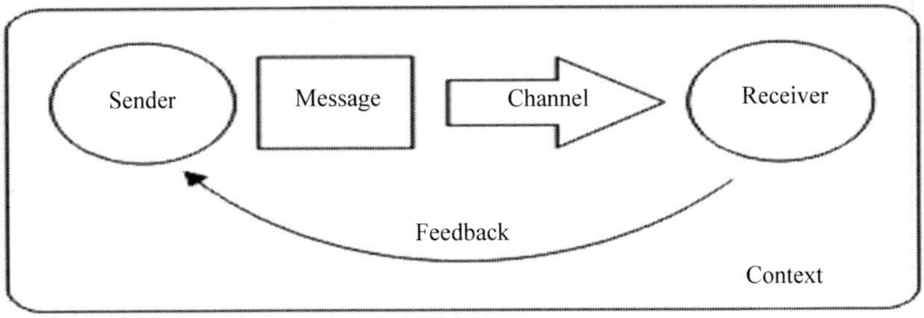

Figure 3.1: Diagrammatic presentation of the communication process.

Elements of Communication Process

In the communication process, there are elements such as source, message, channel, treatment, receiver, feedback, etc. (Figure 3.2).

Communicator: In a communication process the originator of information or message is called the communicator. She/he is the person who starts the process. The communicator sends the message through a channel to reach the receiver or communiqué. She /he gives expression to a message intended to reach an audience to yield the desired response. It is necessary that the communicator should hold the confidence of his audience for effective communication. A good communicator should possess the following characteristics:

- Clear thinking of the message which should be specific with a specific purpose;
- Utilize an appropriate medium for communication;
- Adapted to the occasion and interest of the audience;
- Uses messages in words or pictures to create an accurate image in the mind of the receiver;
- Prepares a plan for selecting, treating the message and expressing the message through a proper channel; and
- Must bc a good listener.

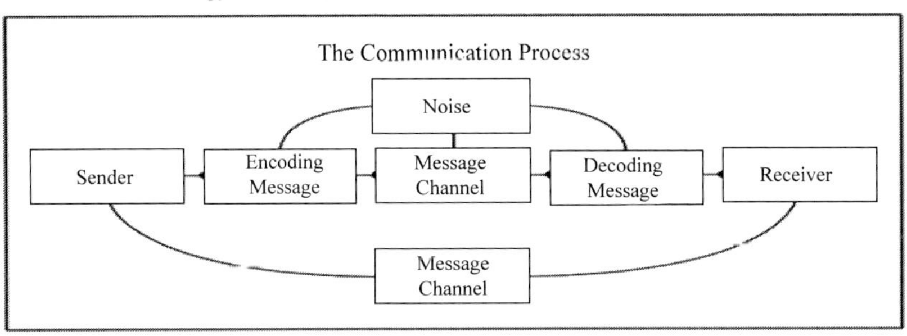

Figure 3.2: Elements of communication.

The message or content: A message is the information communicated from a source to the audience, which they receive, understand, accept, and act upon. The message must be relevant, interesting, useful, profitable, credible, and complete in all respects. A message may consist of statements of scientific facts about developmental programs which will be beneficial to the receiver. For example, an agricultural message may consist of statements of scientific facts about agriculture, a description of action required by individuals, groups, or committees with reasons why certain kinds of action should be taken to achieve a good result.

Channels: A channel is required to convey the message by the sender to the receiver.

The communicator in contact with his audience requires treatment of the message for presentation. The treatment makes the message clear and understandable. While treating the message real objects, models, specimens, photographs, charts, etc. become very useful. Many obstructions referred to as "noise" can prevent the message from being heard by or carried over clearly to the audience. The message can be in written or spoken form. It can be channeled through individual contact, personal communication, or through mass media like Radio/ TV/Internet, etc. The message can be discussed in meetings. The channel should be carefully chosen to fit the requirements of the audience. Written materials cannot be read by illiterate people and all people cannot attend the meetings. Many people may not have access to both print and electronic media.

Audience/Receiver: An audience or a communiqué is the intended receiver or user of the message. It is the intended respondent who can gain economically, socially, or in other ways by responding to the message in particular ways. The more homogeneous an audience, the greater the chances of successful communication. The audience could be potential, available, and active.

Audience Response/Feedback

This is the last element in the communication process. The audience responds to the message received and puts it into action. The feedback of the audience is required to fine-tune the message or to understand the impact of the message. The five elements of message viz., communicator, message, channel, treatment, and audience are intended to be considered for attaining envisaged action by the intending audience. Action taken by the intended audience that can be attributed to a given communication may be assumed to be a result of the degree to which these elements have been effective. After receiving a message, it may be generally expected that audience response will be widely varied in the following ways.

- Understanding vs. Knowledge
- Acceptance vs. Rejection
- Remembering vs. Forgetting
- Mental vs. Physical action
- Right vs. Wrong

Communication Models

In a communication process, the *source* transmits a message through a *channel* to the *receiver*. The message is encoded by the sender which is decoded by the receiver. The source observes the effect and evaluates the impact of the message, which is called *feedback*.

Many models have been proposed in communication to describe the various processes through which a message reaches to receiver from the source. Successful communication will require both the sender and the receiver to share responsibility to make communication smooth, effective, and efficient. A simple diagram indicates the sender (S), sending a message (M) to a receiver (R) (Figure 3.3).

In the above diagram, we cannot know whether the message is properly understood by the receiver or actually received. Sometimes we can learn something by becoming receivers ourselves and watching and listening to reactions to what we have just said or done. Thus communication is most effective when it is a two-way process (Figure 3.4).

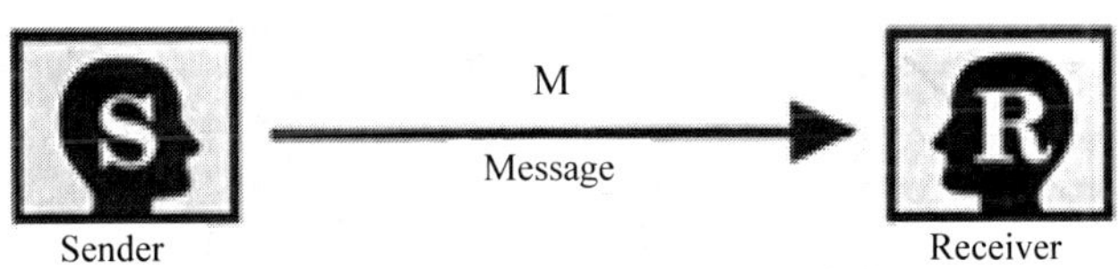

Figure 3.3: One-way communication from the sender to receiver.

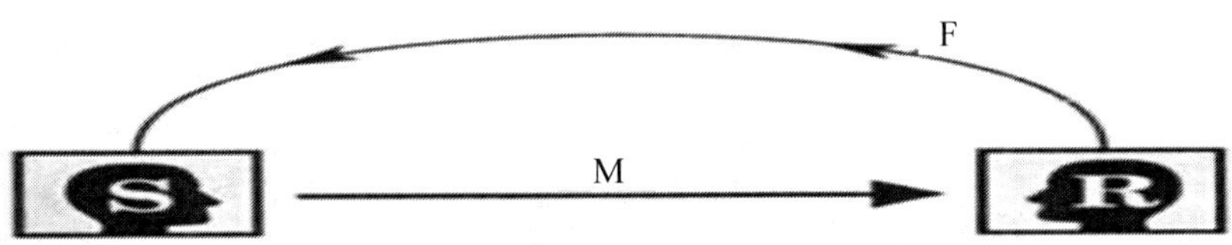

Figure 3.4: Two-way communication.

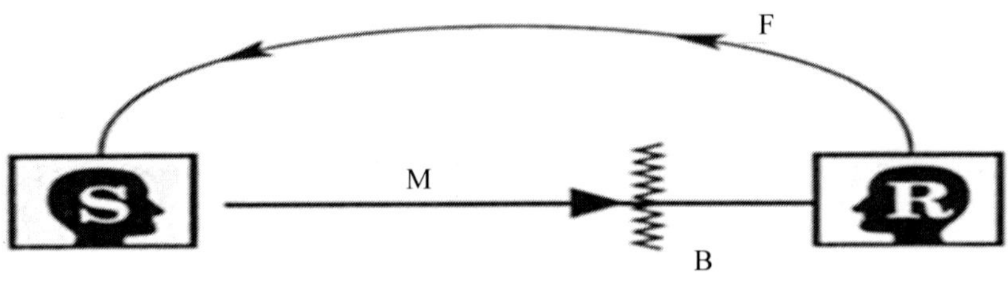

Figure 3.5: Blockage (B) in communication

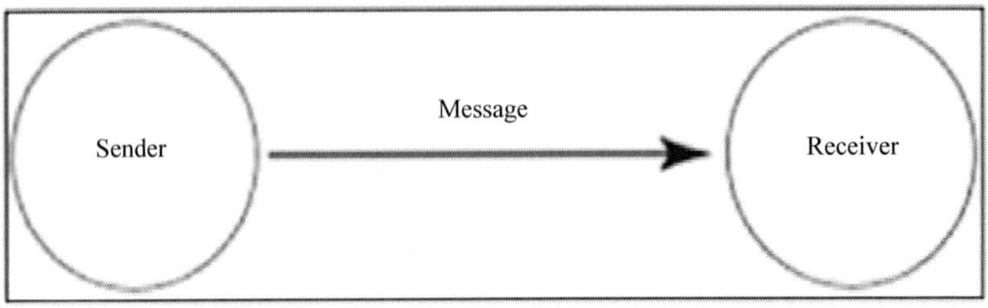

Figure 3.6: Aristotle model

Sometimes our communication interrupts and becomes a miscommunication. There may be some Blockage (B) that appears in the way of our messages. The blockage may be due to some external factor like the noise of a tractor that literally blocks out the sound of our voice (Figure 3.5).

Some important communication models are discussed below.

Aristotle Model

Aristotle, the Greek philosopher, was the first to take the initiative and design a linear model of communication. His model was designed for speaking or oral communication (Figure 3.6). According to him, the speaker plays a key role in communication. This model was widely accepted. However, there have been a few criticisms around this model that there was no concept of feedback because of the one-way model and there may be no concept of communication failure due to noise.

Shannon-Weaver's Model

This model was initially designed to improve technical communication but later used to describe different fields of communication. This model was specially designed to develop effective communication between sender and receiver and identify factors that were affecting the communication process called "noise." Here sender encodes the message and sends it through the channel to the receiver and the receiver has to decode the message before understanding it. The noise was added later due to the failure of the receiver to decode the message due to certain problems. The model is presented in Figure 3.7.

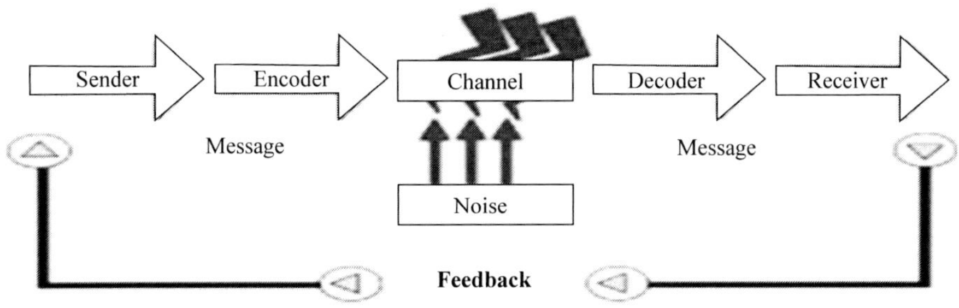

Figure 3.7: Shannon-Weaver's model

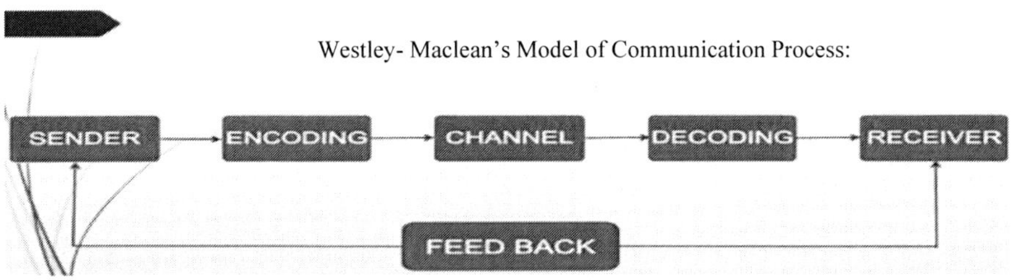

Figure 3.8: Westley-Maclean's model

Welstey and Mclean Model

In 1957, this model was proposed by Bruce Westley and Malcolm Mclean which explains a strong relation between response from surroundings and the process of communication. Here sender encodes the message which passes through a channel and is then decoded by the receiver to make the message clear. This model included the feedback and the model is very descriptive. However, it has some limitations as it does not explain communication which involves multiple messages and complicated messages, and does not include noise in communication (Figure 3.8).

Berlos Model

David Berlo's model (1964) includes certain factors that influence the communication process between two people such as skills, awareness level, cultural system, social system, and attitude. It is believed that the source and the receiver need to be on the same level for effective communication. The difference between Berlo's and Welstey and Mclean's model is that Berlo adds a message to his model (Figure 3.9).

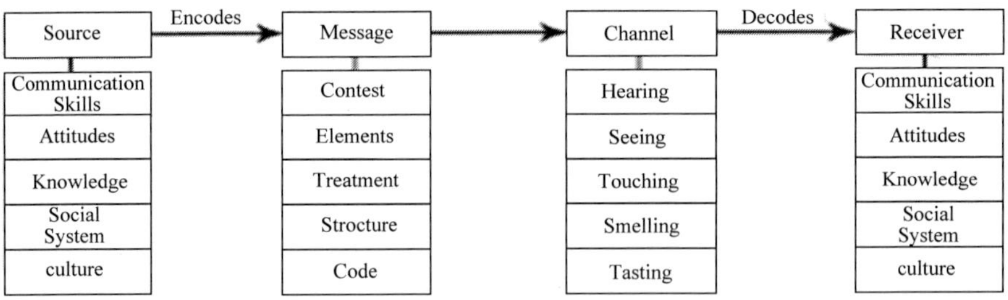

Figure 3.9: Berlo's SMCR model.

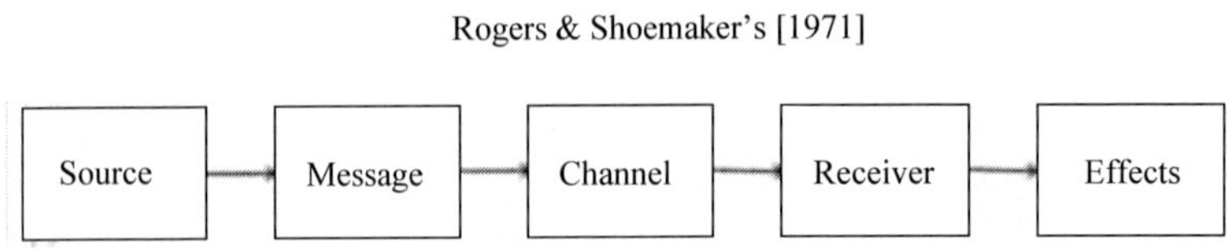

Figure 3.10: SMCRE model.

SMCRE Model

Roger and Shoemaker (1971) developed the S-M-C-R-E model in which the source sends some messages to the receiver through some channels which causes some effects (Figure 3.10).

Osgood–Schramm Model

As per Osgood–Schramm's model (1954) communication is a two-way process between the sender and the receiver. This model involves circular communication between two individuals where both-way communication, rather than one-way communication takes place (Figure 3.11).

A diagrammatic representation of the extension communication system based on the model developed by Leagans (1963) is presented in Figure 3.12.

Importance of Communication in Agriculture

In modern farming, the farmer adopts the latest methods of farming techniques based on science to increase agricultural production. It requires the transfer of new knowledge to the farmers for their adoption in fields. A suitable process of communication is used to transfer such knowledge or technology. It is the agriculture extension that transfers such technologies for well-being of the farmers.

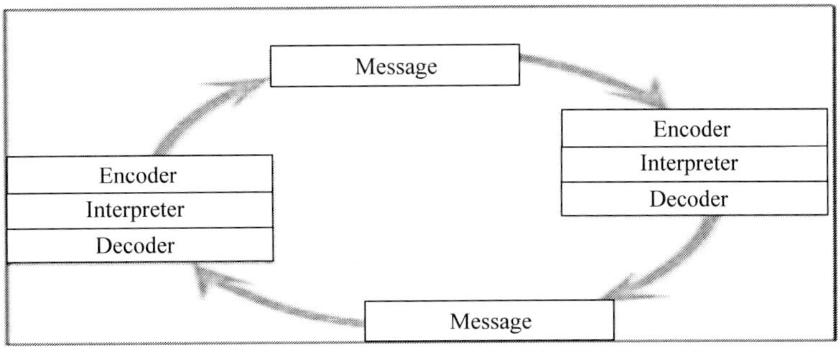

Figure 3.11: Osgood–Schramm model.

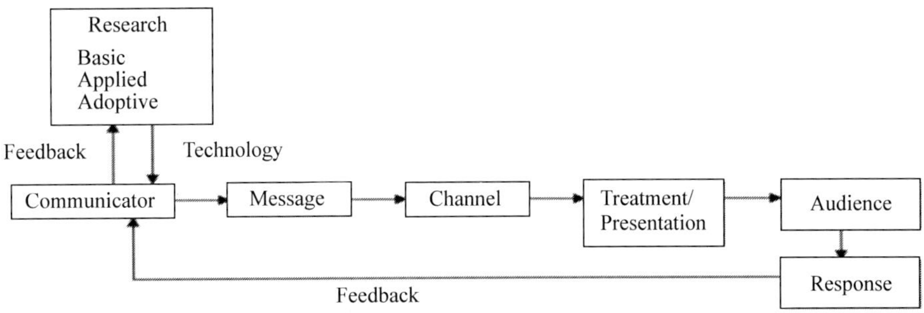

Figure 3.12: Extension communication system.

Scope of Communication in Agriculture

- Newspapers should provide some space for communication of agricultural information and the government should provide some aid to these rural newspapers.
- Booklets and leaflets should be published in which there should be a description of solutions to the problems and the working methods.
- The electronic media like TV, and radio are powerful media for communication of agricultural messages. The government should provide TV sets or FM Radio in rural areas for communication of agricultural messages and Radio/Television coverage should be more on rural development.
- Small films on agricultural technologies and success stories are produced for the farmers. YouTube can release new technologies in smartphones.
- Farmers' fairs and exhibitions can be organized at regular intervals at district and block headquarters to convince many farmers at a place.

Communication Barriers

For an effective communication process, it is needed by the communicator to understand the results through feedback from the receiver(s). There could be many barriers in the communication process as discussed below.

Language barriers: Communication becomes effective when the local language or dialect is used by the communicator. The communicator needs to use local languages and local terms used by the receivers for effective communication. The communicator should avoid jargon and use only simple language that is properly understood by the receiver.

Psychological barriers: The receiver should be in a good mood and willing to receive the message. The message is not understood properly when the receiver is in a pre-occupied mind, in an angry mood, and a stressed condition. The communicator should be quite familiar with the receiver and communicate when the receiver is in a good mood.

Physiological barriers: A good physical state of mind of the communicator and receiver is very important for effective communication. A receiver in a reduced hearing state may not receive the message if the communicator speaks in a low voice or if there is noise. Similarly, the communicator with reduced hearing may not listen to the feedback of the receiver for which the receiver should speak in a loud voice.

Physical barriers: Communication becomes effective when the physical distance is short. Even though modern technologies are capable of limiting the physical barriers in communication, it is desirable to examine the merits and demerits of each communication channel before choosing an effective channel that can avoid physical barriers.

Attitudinal barriers: Behavior and attitudes of both communicator and receiver may cause barriers to communication. Very often, personal conflicts, resistance to change, and fixed mindsets cause barriers to communication. It is the duty of the communicator to manage these barriers for effective communication.

Overcoming the Barriers in Communication

The measures for overcoming the barriers in the communication process through appropriate skills and techniques are discussed below.

Ensuring active listening: Active listening is required by both the communicator and receiver. Active listening skills require fully concentrating on what is being said rather than listening with all senses. There are both verbal and non-verbal ways that convey active listening. Non-verbal communication like smiling, eye contact, head nodding at appropriate times, and avoiding distractions are required for active listening by the receiver. Expressing verbal signs and symptoms of active listening can also be useful for active listening. The receiver can ask a clarifying question to the speaker and the speaker can use different words to repeat the main idea of the speaker to ensure active listening.

Use simple language: The communicator should use simple language and terminology that is understood by the receivers without any difficulty. The speaker may take a pause occasionally and ask questions to draw the attention of the audience and understand whether they are comfortable with the language used.

Reduce distortion of communication: The communicator should give as few details as possible, make specific and clear-cut statements, use visual aids, and ask questions to confirm the understanding of the audience.

Give constructive feedback: The communicator should allow the listener to ask questions to clarify any points. Care should be taken to see that the feedback is constructive in nature. However, negative feedback may also be accepted politely.

Problems in Communication

Communication problems originate from the sender or receiver. The usual problems are lack of sensitivity with the communicator, lack of communication skills, lack of knowledge, overloading of information, emotional interference, and noise.

Problems may also arise at the communicator level, message level, and at the level of the receiver. The problems with the communicator are an ineffective environment, disorganized efforts to communicate, the standard of correctness, the standard of social responsibilities, cultural value and social organization, inaccurate symbols, and wrong concept of the communication process.

At the message level, many obstructions (noise) can enter the interpretation level. "Noise" emerges from many sources and reasons which affect the interpretation of the message. The problems at the message level are wrong handling of the channel, wrong selection of channel, physical distraction, or inadequate use of channels in parallel.

The problems arising at the level of the receiver are the attention of the listener, the problem of participation and involvement, the problem of homogeneity, and the attitude of the audience toward the communicator. The communicator should develop a favorable attitude to solve the problem.

Participatory Communication in Extension

Effective communication depends upon the perception of the individuals involved. It is said "you see, what you want to see" and "you hear, what you want to hear." The perception differs from individual to individual as per culture, family background, and level of education. The difference in perception between the extension worker and the farmer(s) makes the communication ineffective. It often happens that either of them gets a different signal from the other in the process of communication.

The extension worker should participate wholly with the farmer without maintaining a distance and ensure the participation of the farmers. He must realize that "Farmer Problem is His Problem." He should talk at the level of the farmer as a family member and should create such a situation so that the farmer does not feel that he is bound. One-way communication does not lead to any development for which the communication should be made two-way. Careful use of extension methods and channels could trigger participatory communication.

References

Anderson, W.A. 1969. Disaster warning and communication procesd in communication. Journal of Communication.19. no.2(1969) 92-104.

Ban, Van Den, A.B. and Hawkins H.S. 1998. CBS Publishers and Distribution, New Delhi.

Benor, D. and Harrison, J.Q. 1977. Agricultural Extension- Training and Visit System, World Bank, Washington.

Benor. D. 1984. T&V: Putting intentions into practice. INTERPAKS Imer• change. 1 (No.2), 1-2.

Berlo, D.K. 1960. The process of communication. Holt, Rinehart and Winston, New York.

Chandrakandan, K., Senthil Kumar, M. and Swatilakshmi, P.S. 2005. Extension Education What? And What Not? RBSA Publishers. Jaipur.

Dass, R. 1981. Appropriate Technology – Percepts and Practices. Vintage Press Inc., New York.

Dhama, O.P. and Bhatnagar, O.P. 1987. Education and communication for Development. Oxford and IBH Publishing Co. Pvt. Ltd.

Dhama, O.P. and Bhatnagar, O.P. Education and Communication for Development. 2007. Oxford and IBH Pub.Co.Pvt. Ltd ISPA - Indo Swiss Project Andhra Pradesh (Year), Learning Material, Directorate of Animal Husbandry, Hyderabad.

Kelsey, L.D. and Hearn, C.C. 1954. Cooperative Extension Work, Cornell University Press, Itheca, New York.

Leagans, J.P. 1961. "The Communication Process." In Extension Education in Community Development. New Delhi: Directorate of Extension, Ministry of Food and Agriculture, Government of India.

Mclean S. The basics of interpersonal communication, 2005.

Niels 1988. Extension Science - Information Systems in Agricultural Development, Cambridge University Press, Cambridge, UK.

Orivel. F. 1983. The impact of agricultural extension: a review of the literature. In Perraton and others (Eds.), Basic education and agricultural extension. Washington, D.C.: The World Bank.

Ray, G.L. 1991. Extension Communication and Management. Kalyani Publishers, Kolkata.
Rayudu, C.S. 1997. Communication. Himalaya Publishing House, Mumbai.
Rogers, E. M. 1995. Diffusion of Innovation. The Free Press, New York (1995).
Rogers, E.M and Shoemakers, F.F. 1971. Communication and innovation: A Cross-Cultural approach. 2nd Edition, The Free Press, New York.
Singh, Ranjit. 2003. Extension Education, Jaswant Printers, Ludhiyana.

Suggested Reading

Van den Ban, A.W and Hawkins, H.W. 1996. Agricultural Extension, 2nd Edition, Blackwell Science, London.

4

Extension Teaching Methods

Introduction

A teacher uses different aids in addition to lectures for effective teaching. Similarly, a trainer uses different training methods to make learning effective. In Agricultural Extension the extension worker as a teacher has to design and use appropriate teaching methods depending upon the group to make the teaching effective. According to Leagans (1961) "extension teaching methods are the basic and proven devices used to create situations in which communication can take place between the instructor and the learner." Extension methods are powerful tools for extension workers to effectively communicate certain messages. Some studies suggest that the adoption of innovations by the farmers requires different teaching methods at different stages of learning. These methods are required for imparting new knowledge and skills to the farmers to adopt new practices for their increased production and income. The choice of extension teaching methods depends on the target audience, location, and availability of time for teaching.

Classification of Extension Teaching Methods

Extension teaching methods are classified according to use, form, and function as shown in Table 4.1.

Classification According to Use

The extension methods could be individual or interpersonal, group contact, or mass contact as shown in Figure 4.1.

Individual contact methods: Individual methods provide face-to-face contact with individuals. Those techniques are very effective in teaching new technologies and creating goodwill among farmers and extension people. When few people are to be contacted and they are at a close distance, the individual methods are effective. The extension worker contacts directly to an individual and gets his feedback through different means like personal visits, telephones, SMS, etc. The individual contact method is useful for rapport building and easy to obtain feedback. But the method is time-consuming and quite expensive.

Table 4.1: Classification of Extension Methods

Particulars	Methods	Examples
According to use	Individual	Farm and home visits, telephone calls, personal contact, office calls, personal letters, minikits Trial, SMS, WhatsApp
	Group	Demonstration, meetings, group discussion conference, seminar, tours, field trips
	Mass media	Farm circulations (bulletins, newsletters, pamphlets, bulletins), campaigns, exhibitions, fairs, Television, Radio, Films, Videos
According to form	Written	Letters, leaflets, bulletins
	Visuals	Demonstration, posters, exhibits, slides, charts
	Spoken	Farm and home visit, official calls, telephone, general meeting, Radio, Audio cassettes
	Audio-visual	Television, Video, PPT
According to function	Telling	Lecture, meetings, Audio lesson, farm and home visit, Radio, Extension talk
	Showing	Exhibition, tours, motion pictures, video text, Internet
	Doing	Practical work, workshop, demonstration, Do it yourself

Figure 4.1: Extension methods as per use.

Group contact methods: The farmers are contacted by extension workers in a small (15-25) or large group (50-100) which are formed around a common interest. This method provides face-to-face contact with the people and gives an opportunity for the exchange of new ideas, decisions on problems, and technical solutions. Cross-learning takes place between individuals. The usual group contact methods are Field demonstrations (Method and result), Group meetings, Lecture, Conferences, Tours, Farmer Schools, Workshop, Symposium, Brainstorming, Debates, Panel discussion, Field Days, Field Trips, Seminar, Roleplay, Informal group discussion, etc. This method is cost-effective and provides opportunities for interaction and feedback. Group dynamics is necessary to make the method effective.

Mass contact methods: When a large number of people stationed at different places are to be contacted in a short time, mass contact methods are used. These methods provide scope for quick learning and creating awareness about new ideas. The mass contact methods generally used by the extension worker are Printed materials (Newsletters, bulletin, Leaflet, Pamphlet), Radio talks, Television programs, Posters, Campaigns, Exhibitions, Motion pictures, Drama, etc. This method is effective in communicating with a large audience with least investment. However, there is little scope for getting feedback.

Classification According to Form

According to form, the extension methods could be in written, spoken, visual, and audio-visuals.

Written: Newspaper, personal letters, circular letters, leaflets, folders, bulletins, pamphlets, bulletins, Newsletters, etc.

Spoken: Farm and home visits, meetings, official calls, and Tape recording, etc.

Visual: Demonstration, posters, photographs, blackboard, result demonstration, slide projectors, OHP, etc.

The visual methods can be further classified as projected, non-projected, and audio-visuals.

Projected aids: Movie films, Movie-strips, Slides, overhead projector (OHP), opaque projector cloth, LCD projector, etc. are projected aids used in the extension.

Non-projected visible aids: Images and flash images, posters, wall information sheets, flashcards, strip charts, magnetic boards, samples, cartoon pictures, etc.

Audio Visual: Television, motion pictures, multimedia projector, PPT, etc.

Common Extension Methods in Use

Commonly used extension methods by the extension works are discussed below.

Individual Contact Methods

It is a face-to-face contact at home or on the farm for discussion and transfer of need-based technology (Figure 4.2). During face-to-face contact the farmer gets the scope to discuss his problems freely with the extension worker and gets individual attention. The personal influence of the extension worker is useful for teaching new skills and helping a farmer at his place in critical decisions. During personal contact, either in the field or at the office of the extension worker, the farmer listens to the advice of the extension worker and gets individual attention. Some common individual methods are discussed below.

Figure 4.2: Individual contact.

Farm visits: A farm visit is a common form of personal contact between the extension worker and the farmers at the farm of the latter (Figure 4.3). During this visit, the extension worker becomes quite familiar with the farmers and gives specific advice on critical issues and solving the field problems on the spot. Apart from transferring new technologies, the extension worker can study the impact of the technologies on field conditions and get feedback from the farmer.

Figure 4.3: Field visit.

In case of an emergency, of urgency, the extension worker can make instant visits.

However, the farmer's visit should be planned and known to the farmer in advance for effective learning. Before the field visit, the purpose of the visit, the field/crops to be visited, and the names of the farmers to be visited should be decided and it must be planned to fit into other local extension activities. The extension worker should meet the farmer at his workplace and at a time convenient to the farmer. To make the field visit effective, a checklist can be drawn in advance by the extension worker. The purpose of any farm visit will be defeated if its content and conclusion are not recorded and no follow-up action is taken.

Office calls: The farmer should be allowed to visit him/her in the office in case of urgency. Such office visits are less time-consuming and convenient for both. During such official calls, the extension worker should be courteous and try to dispose of the farmer as quickly as possible. He should not feel it is a burden and create a good environment to make the visit of the farmers should be cordial and purposeful.

Letters: Correspondence with a farmer through letter is a cheaper method of communication. Letters serve as a follow-up inquiry after an extension worker's farm visit, or sent when a farmer is unable to make a personal office visit. The extension worker should write a letter in the local language with important messages in brief. The letters should be short, precise and clearly written.

Telephone calls: When facilities exist, communication by telephone is time-saving and it becomes easy to get feedback. The telephonic discussion may be a short one and communicate definite messages. It is better if the extension worker notes down the important points to communicate. The farmer notes down the points communicated to him so that the communication gap can be avoided. In case of urgency, SMS or WhatsApp messages can be communicated to the farmer. Video calls can be made by the extension worker with the farmer, if both are using smartphones. Kissan call centers are now available for individual contact (Figure 4.4).

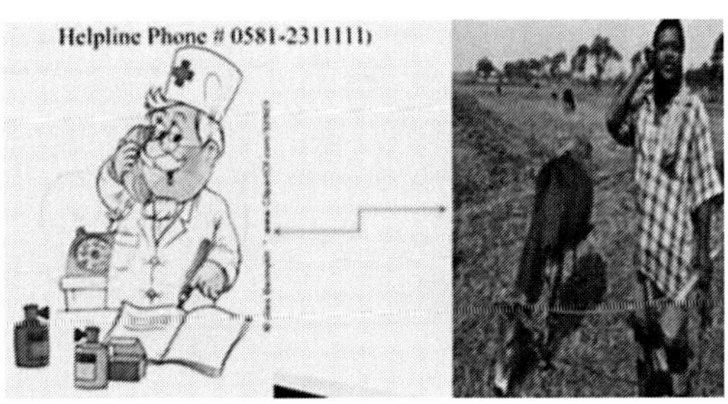

Figure 4.4: Kissan call center.

Group Contact Methods

Group extension methods are very effective as cross-learning occurs in a group. It allows greater extension coverage within a limited time. Group extension offers a more reflective learning environment in which the farmer can listen, discuss and decide upon his involvement in the extension activity. Individual farmers can gain greater self-confidence in a group extension by joining others to discuss new ideas and try out new practices. Group extension is very effective to deal with common problems. The important group discussion methods are discussed below (Figure 4.5).

Figure 4.5: Group contact method.

Group meetings/discussions: In a group meeting the extension worker and farmers sit together at a common place to discuss important messages. The extension worker can discuss new government policies, projects, etc., and answer all questions posed by the farmers in a group meeting. Specific topics of interest such as horticulture, beekeeping, or dairy farming are presented and discussed in detail by the extension worker(s). In general meetings, men, women, and young and old participants take part in discussions. However, in a specific meeting, only men, women, and vulnerable people can join separately and take part in discussions relating to their problems. The group meetings should be formal or informal (Figure 4.6). Many people should be encouraged to express points of view and ask questions in the meeting. Discussions may be completely open and unstructured, or based on a prepared agenda of discussion points.

Figure 4.6: A group meeting.

Proper planning is required in advance to convene a meeting indicating the place, time, and purpose of calling the meeting. The meeting place should be well-known, accessible and comfortable. It is better to prepare in advance a checklist for use in a meeting. After the discussion is over the proceedings of the meeting should be recorded and read before the farmers.

Field Demonstrations

Field demonstrations are laid out in the field based on the principle of "Seeing is believing." A practical demonstration is an effective method to convince the farmers about a new idea or practice. It gives scope for farmers to know the differences between a recommended new crop practice and traditional practices which they adopt. There are two types of demonstrations, such as method demonstration and result demonstration.

Method demonstration: In a method demonstration the practices are shown step by step to convince the farmers about the comparative advantage of new practices. The main advantage of the method demonstration is that the extension worker can explain required farming skills to a large number of people on the site by example. As farmers participate in the program, there is a greater chance of getting benefits from the technologies demonstrated. Field Days can be organized during the harvesting of the crop to show the benefits to a large number of farmers (Figure 4.7).

Figure 4.7: Method demonstration.

Result demonstration: A result demonstration is laid out in a farmer's field along with a control plot side by side to prove the superiority of the new practice over the old practice in terms of good yield. "Seeing is believing" is more appropriate to a result demonstration. The farmers get convinced easily by seeing the tangible results of a new practice recommended by the extension service. It creates confidence among the farmers to adopt the new practice (Figure 4.8).

Figure 4.8: Result demonstration.

In agricultural extension, both result and method demonstrations are very useful and need careful planning and execution. Although the two demonstrations differ somewhat in their purposes, they require common planning and execution.

Field days: Field days are organized at the demonstration site at the time of harvest of the crop to display the results to convince a large number of farmers of the superiority of the new practice over the old practice with crop cutting results. Experimental stations or other government centers or even the result demonstrations can be used for field days (Figure 4.9).

Figure 4.9: Field days.

A small to a large group can be present on field days. The role of the extension worker is to support the farmer on whose land it is being held and to explain the other farmers about the new idea or practice.

Field Trials: Adaptive or mini-kit trial is a method of determining the suitability of a new practice or crop variety in farmer situation. These are laid out in a farmer field by adopting proper design.

Tours/Exposure visits: During the study tour, the farmers are exposed to new innovations and practices adopted in other areas. A tour is organized to a series of field demonstrations on different farms, or at different centers, which can convince the farmers by seeing the result of better practices themselves (Figure 4.10). It offers a chance for the farmers to see how other farmers cultivate their land, and to exchange ideas and experiences with them. The site to be visited may be of similar location of the farmers.

Figure 4.10: Exposure visit.

Farmers' Field School (FFS): The Farmer Field School is a farmers' learning platform for farmers which provides them scope for observation and experimentation. Initially FFS was organized to convince the farmers to learn about Integrated Pest Management Practices. Regular sessions from planting till harvest are organized at suitable intervals to teach the farmers on the dynamics of the crop's ecosystem and management options. In this cyclical learning system, farmers increase their knowledge that enables them to make their own decisions in crop management.

The Integrated Pest Management was very successful in FFS. Many new methods such as System of Rice Intensification (SRI), Organic farming, Zero Budget Natural Farming (ZBNF), etc. are included in the FFS program.

Concept of FFS

The Farmer Field School (FFS) approach represents a radical departure from earlier programs, in which farmers were expected to adopt generalized recommendations that had been formulated by specialists from outside the community (Figure 4.11). FFS consists of a group of people with common interests who get together on a regular basis to study the "how and why" of a particular topic under field situations. The farmers go through a subject-oriented, discovery-based totally training that allows them to come to be subject experts and be capable of grow a wholesome crop.

Figure 4.11: Farmers field school

- The topics in FFS can vary substantially from IPM, organic agriculture, animal husbandry, and soil husbandry to income-generating activities along with handicrafts.

- FFS is a participatory extension methodology that acknowledges the need to involve farmers in technology development and its transfer.
- FFS emphasizes constructing on the farmers' potential to test and draw conclusions and it empowers farmers to improve their socio-monetary situations.
- It operates with the precept of non-formal schooling (non-formal learner-focused educational manner) and a maximum of the consultation and contents are based on adult education. As the name suggests FFS are the schools for the farmers outside the classrooms and on the farms. FFS is basically "school without walls."
- In this process, farmers are central in the process of technology development.
- FFS emphasizes capacity building and empowerment of farmers to experiment and draw conclusions to improve their socio-economic conditions.
- FFS however, is particularly adapted to the field's study, where specific hands-on management skills and conceptual understanding are required to be given by the facilitators.
- It seeks to empower people to solve their problems actively by fostering participation, interaction, dialogue, joint decision-making, self-confidence, and self-determination.
- Farmers learn by doing after carrying out various activities related to a selected farming technology and through constant observation of the technology performance in the field with their practical knowledge acquired from generation-long experience.
- Field operations in FFS plots take place in parallel with the operations in farmers' fields.
- The FFS is a tool for capacity building of farmers and NGOs to make them better decision-makers in promoting sustainable use of resources at the cropping, farming, and watershed system levels.
- In FFS, farmers learn about sustainable agricultural practices on the field. The fields belong to an experiment station or a community. There, farmers meet regularly for the duration of an entire cropping season. They learn through regular field observations from pre-planting to harvest.

Mass Media Methods

Mass communication covers extension publications such as print media and electronic media to reach a large number of farmers stationed at different locations. Extension publications containing the farm information are used for communication with a large number of literate people. These publications may be used singly or in combination with other extension methods for communicating general & specific technology or practice. Leaflets, Pamphlets, Folders, Bulletins, Journals, Newsletter, and Circular letters are some of the important print media. Electronic media covers audio (radio, cassettes), and pictorial (television, film, and video). The electronic media are more powerful than the print media as they explain the practice step by step with appropriate visuals. Although mass media are the cheapest source of wider communication, they cannot be the substitute for extension workers.

Some of the important mass media are discussed below.

Electronic Media

Radio/FM Radio: Battery-operated or electric radios are still used in rural areas. Due to its wider coverage, information can reach households directly and instantly throughout the covered region. Radio is preferred as a convenient medium of communication in times of natural calamities. In such a situation, radio plays a significant role as no other electronic communication is feasible due to electricity failure. Rural people without any educational background can take advantage of this medium and build up adequate knowledge and skills. The broadcast may be heard by the farmer when he is engaged in farming or household work. The farm home broadcasts inform the farming community about the latest scientific techniques for increasing production in all important farm enterprises and in the non-farm sector. There are some difficulties like the cost of the battery, want of repair facilities and specific coverage time, etc. in rural areas.

Audio cassettes: Audio cassettes are prepared on definite subjects of communication which can be used in cassette players. The cassettes can be repeatedly heard to clarify doubts. These cassettes are also used by the extension workers as teaching aids in training. Now YouTube has become very popular in communicating technical messages with visuals.

Film: Film shows are regarded as a source of entertainment and communication in rural areas. It is an audio-visual aid for forceful learning as the farmers can hear and see the technical messages simultaneously for better understanding. The close-up shots and slow motion create interest among the audience for a better understanding of the subjects. The success stories collected can influence teaching that cannot normally be seen directly. However, it is not a popular extension method as it requires higher cost and costly equipment.

Television and video: Television is an effective medium for the transmission of messages quickly in live form. Specific programs are prepared as per the current need before the actual telecast. Government programs are telecasted on their channels at a fixed time.

Video combines the visuals with an audio system for viewing on a monitor or television set or on a computer screen. Rerecording is feasible in video recording. Video is more suitable than films as video programs can be prepared speedily with multiple copies at a lower cost. They are also very easy to distribute in the form of video cassettes.

Printed Media

Printed media can be looked at frequently and can be referred to again and again. The print media are more authenticated than other media. Printed media used in extension includes posters, leaflets, circular letters, newspapers, magazines, etc.

Posters: Posters use broad messages in bold letters which can be seen by the farmers for a long time. They are used for publicizing future events and messages in a common place to attract the audience.

Leaflets: They are printed materials on a small piece of paper for wider circulation of messages to a large number of farmers.

Circular letters: These are used to publicize results of extension and important messages as reference material for many farmers who cannot attend them may still benefit.

Pamphlet: A pamphlet is an unbound sheet of paper with the printing messages on both sides printed in colors with motion photos, giving full statistics approximately on a topic in greater length than in folders or leaflets.

Bulletin: Bulletins are issued as small booklets of about 20 pages with the objective of giving whole information that the user can follow under his/her personal situation.

Booklet: Booklets are small books that contain information in 20-50 pages on important technologies or messages.

Newspapers: Newspapers covey agricultural news items and keep a page for agricultural-related topics for creating awareness among the farmers who can get benefits from these papers.

Exhibits and Displays

An exhibition is a systematic display of models, live specimens, charts, photographs, pictures, posters, information, etc. in a public place for awareness generation and interest of the audience. They are held at many different levels, ranging from village to international. Exhibitions and fairs are used for awarding prizes to the selected farmers for their outstanding performance and provide scope for interaction of scientists and farmers.

Campaigns: Several media are used in a campaign mode to convey messages in a coordinated way for a limited period. It aims at a particular extension objective. It is necessary to make all arrangements for the film shows, group meetings, demonstrations, etc. along with wide publicity before the visit of staff for the campaign. Campaigns are very successful in locations where farmers face common problems and the possible solutions are discussed in the campaign with the use of appropriate media.

Traditional media: In rural areas there are still some traditional media like street songs, folk dances, dramas, plays, etc. to serve as a source of entertainment through which important messages are communicated to people. These media are used when the level of literacy is low and extension workers are involved in including agricultural messages in such media.

ICT-based Agriculture Extension

National policy for farmers (2007) indicated that the ability of ICT could be harnessed through establishing *gyan chaupels* (information centers) in villages and the Common Service Centers (CSCs) by state governments under the Ministry of Communication & IT, besides evolving private initiative programs for inclusive broad-based development. Providing last-mile connectivity could be facilitated with the assistance of technologies including broadband net, network radio, or internet-cellular telephone synergies (NPFF 2007).

Some of the important ICT systems in agricultural extension are AGRISNET, DASNET, e-Kishan, Rice Knowledge Management Portal (RKMP), Agropedia, i-Kishan, Village Knowledge Center (VKC), Kisan Call Center, Mobile Advisory Services by KVKs, and Digital Green, etc.

Internet: The Internet often referred to as 'the network of networks' is a global network of computers that are linked together by cables and telephone lines making communication possible. It stores a wide range of information that can be accessed with a click. Many people, cutting across geographical boundaries, can interact through it, using as well as providing the information it contains.

World Wide Web: Required information can be searched by agriculture scientists, students, extension workers farmers, and traders in a very short time, with the browsing of the World Wide Web (WWW).

E-mail: E-mail has replaced paperwork like letters and faxes and communicates information within a short time. However, in rural areas, the use of e-mail is very limited due to the non-availability of connectivity, the lack of technical personnel, and facilities.

Kisan Call Center: The Department of Agriculture, Government of India, has launched state-wide Kisan Call Center to answer the queries and questions asked by farmers.

Selection of Methods

Selection of appropriate method depends upon many factors outlined below:

The Audience

Human beings range significantly in their understanding, attitudes, talents, function within the "diffusion process", and in the "adoption categories", their educational training, age, and socio-cultural patterns. Some of them are looking for progressive changes, and a few are slow in the adoption process. These persons and their collective differences influence the teaching technique. Usually, people with little or no education and low incomes may respond to personal visits and result demonstrations. The higher educated and the extra innovative factors of the population usually respond well to methods like group meetings and discussions, exhibits, and written materials.

The size of the audience is also a factor influencing the choice of extension methods. Group discussion is effective when the group strength is 20-30. Method demonstration is suitable for a small group of farmers and a large audience can be taught by lecture method.

The Teaching Objective (Or Nature of Change Aimed At)

Mass media are used to influence a large audience to create awareness or pass some information, while individual contact methods are used to influence small groups for maximization of communication. Group discussions are held or village leaders are engaged when the purpose is to change attitudes or arrive at a consensus of opinion. Methods like method demonstration are used for individual skill teaching in the field.

The Subject Matter

In the case of simple technologies and familiar practices, methods like news articles, radio talks, or circular letters are useful. For complex and unfamiliar messages and practices, personal contact, written instructions, and audio-visuals would be effective.

The Stage of Development of Extension

In order to convince the farmers about a new practice, result demonstration will be an appropriate method and in the case of a credible extension system result demonstration is required. In such cases, it is better to make use of local leaders or local illustrations for communication of new messages.

Size of Extension Staff

The scope for direct or personal contact methods is greater when the extension workers are adequately available and funds are available.

Availability of Communication Media

Availability of media such as newspapers, telephones, radio, internet, etc., will also have a direct bearing on the extent to which these methods are used.

The Relative Cost

The cost-effectiveness of individual methods is to be considered while selecting a method of communication.

An Extension Worker's Familiarity

Familiarity with and skill in the use of the several extension methods will also influence his choice and use of the methods.

Combination of (Media Mix) Teaching Methods

Audio-Visual Aids

To maintain a high level of understanding and interest and to encourage greater participation and to use them in all levels of learning audio-visual aids are to be used.

Guidelines for Using AV Aids

Planning AV aids: Audio-visual aids, including overhead transparencies, flip charts, slides, chalkboards (and occasionally film and video), etc. can supply added impact and clarity to the presentation of extension workers. The use of A/V aids can visually reinforce the points made verbally in the presentation, summarize the points made, as well as those that have already been made, and visually clarify important concepts and analogies. Audio-visual aids not only help the target group follow the extension presentation, but they also provide support to the presenter by helping to keep him/her on track. It should be remembered that the A.V. aids are intended to add a visual dimension to the oral presentation for reinforcement (Table 4.1).

Table 4.1: Guidelines to be Followed while Planning the AV Aids

Do's	**Don'ts**
Use them in sequencing the content	Do not project the printed text and summarize the key points instead.
Use them for the interpretation of statistics visually with the help of graphs and charts in support of the presentation	Do not only read out the contents written in the AV aids. Rather try to explain the contents for understanding of the audience.
Show them to support the statements	Do not use copies of your transparencies as hand-outs. If required, distribute the hand-outs and written texts after the presentation.
Visually present to add visual clarity to your concepts and ideas	Do not use charts, graphs, or tables that contain more information than you want to provide. Use simple aids and explain them properly.

Table 4.2: Type of AV Aids and their Use

Type	**Use**
Overhead Transparencies	Used for displaying charts and graphs. High visibility for large groups.
Flip Charts	Used for small groups (under 20). for charting and graphing. Showing summaries and sequences. Listing, recording, outlining, developing charts.
Blackboards/Whiteboards	Spontaneous presentations. Listing/recording. When no other media is available
Slides	Displaying high-quality artwork. Displaying a series of photos. Demonstrating- a "hands-on" process. Creating a mood, or inspiring
Videotapes/films	Demonstrating a process. Creating a mood, or inspiring. Presenting testimonials. Supplementing presentation ideas.
Hand-outs	Providing background information. "Sending home" your ideas. A reminder of follow-up activities. Providing a record of the presentation. Weighting a variety of alternatives. Clarifying abstract or complex topics
OHP	Overhead transparencies are inexpensive, easily prepared, and highly effective visual aids for target groups of five or more people. Extension workers are using PPT in PowerPoint.
PPT	PPT is a very effective visual aid. It is displayed by using the computer projected on a screen by the LCD projector.

Choosing appropriate Media: You may wish to combine different types of media in a single presentation. The media used most often are overhead projection transparencies, flip charts, slides, blackboards, and hand-outs. You may occasionally use videotapes and/or films.

Plan to practice using several different media in this course. You will find that practicing will increase your comfort level, and therefore afford you the flexibility of choosing the most appropriate medium. In the future, use the guidelines to help you choose the media best suited to your presentation (Table 4.2).

References

Legans J.P. (1961). Characteristics of Teaching and Learning in Extension Education, Extension Teaching Methods, The Communication Process and Programme Planning to Meet People's Needs in Extension Education in Community Development. New Delhi: Directorate of Extension, Govt. of India., New Delhi. National Policy for Farmers, NPFP, 2005.

Suggested Reading

A.P. Samanta, Samanta, R.K. 1991. Development Communication for Agriculture, (ED), B.P. Publishing Corporation, New Delhi.

Agricultural Support Assist Project (ASAP) Japan International Cooperation Agency (2008): Agricultural Extension Methodology: A reference manual.

Dahama, O.P. and Bhatnagar, O.P. 1987. Education and Communication for Development, Oxford and IBH publishing Co., New Delhi.

Directorate of Extension Govt. of India, 1961. Extension Education in Community Development, DOE, Ministry of Food and Agriculture, GOI, New Delhi.

Food and Agriculture Organisation: Extension Methods: https://www.//fao.org.

R.K. Samanta 1989. Management in Agriculture and Rural Development, UDH Publishing Co., New Delhi.

Ray, G.L. 1991. Extension Communication and Management, Nayaprakash, Calcutta A.A 1987. Extn. Education, Bapatla, Sree Laicshrni Press.

TNAU, Coimbatore 2012. Factors influencing in selection of extension teaching methods.

Van den Ban, A.W. and Hawkins, H.S. 1998. Agricultural Extension; New Delhi: CBS Publishers & Distributors, ND.

Yella Reddy, N. 1998. Audio-Visual Aids for Teaching, Training and Extension. Haritha Publishing House, Hyderabad.

5

Adoption and Diffusion of Innovations

Introduction

Innovation as a new idea, method, or object may be developed by research or by a farmer. Diffusion is a special type of communication on innovations. The proper use of innovation is called adoption. Communication is different from diffusion in the sense that communication covers all types of messages whereas diffusion is concerned with a message of new ideas. The extension workers need to understand the meaning of diffusion, adoption processes, innovation, etc. by which farmers accept new ideas. Adoption is a sequence of thoughts and actions which an individual goes through before s/he finally adopts a new idea. The adoption process is a mental process for passing the new idea to final adoption. In the diffusion process, the new idea spreads from its source of invention or creation to its ultimate users or adopters. Perception is an activity through which an individual becomes aware of objects around oneself and of events taking place.

Elements in the Diffusion of Innovations

The four main elements in the diffusion of innovations are innovation, communication channels, time, and social systems.

The Innovation

Innovation is a new idea expressed in terms of knowledge, persuasion of decision for adoption. The perceived attributes of innovations are relative advantage, compatibility, complexity, trialability, and observability.

Relative advantage: The innovation should have a relative advantage over the existing ideas in terms of economic and social-prestige factors, convenience, and satisfaction. The advantage can be influenced by providing various incentives.

Compatibility: The innovation should be compatible with the socio-cultural values and beliefs of people.

Complexity: When the idea requires complex understanding, it may not be easily adopted by the farmers. This is the reason why simple ideas are promptly adopted by the farmers.

Trialability: Farmers will be inclined to adopt a new idea that they have tried on a small scale first. Trailability imparts less uncertainty to the farmer for adoption as it is possible to learn by doing.

Observability: Once the farmer observes the good result of a new idea and discusses it with others, he will be interested in adopting the innovation.

Communication Channels

The communication channels are interpersonal channels, mass media channels, etc. for the transfer of messages from the sender to the receiver. An appropriate communication channel would be required for the adoption of new ideas.

Time

Time is an important aspect of communication. The time dimension is involved in diffusion in the (i) innovation/ decision process, (ii) innovativeness of an individual or other unit of adoption, and (iii) innovation's rate of adoption in a system.

Social System

The diffusion of innovations will take place within the boundaries of a social system that involves individuals, groups, organizations, etc.

Adoption Process

There are five stages in the adoption process, such as awareness, interest, evaluation, trial, and adoption.

Awareness

At this stage, an individual becomes aware of some new idea such as maize hybrid or a new pesticide. The individual should be made aware of all details of the innovations for taking a decision for its adoption.

Interest

At this stage, an individual will develop an interest in knowing more about the innovation as to how it will be suitable to his requirements. He may ask himself if this might help him increase his income, help him control insects or diseases, or improve farming or home life in some other way.

Evaluation

At this stage, the individual makes mental application of the new idea to the present and anticipated future situations and decides whether or not to try it.

The individual may judge the worth of the innovation and make an assessment of whether the idea is applicable to their own situation, and if applied what would be the result. By evaluating the innovation s/he will be in a position to assess the usefulness of the innovation.

Trial

The individual actually applies the new idea on a small scale in order to determine its utility in their own situation. If s/he decides that the idea has possibilities for him s/he will try it. The trial stage is characterized by small-scale experimental use, and by the need for specific information which deals with the adoption of the innovation. Apparently, individuals need to test a new idea even though they have thought about it for a long time and gathered information concerning it. Trial may be considered as the practical evaluation of the innovation which can provide evidence of the advantages of the innovation.

Adoption

It is the final stage of diffusion as the idea is used with satisfaction. An individual tries the innovation on a small scale as appropriate to his situation on a continuous basis after his satisfaction with its evaluation and trial. However, the adoption process does not follow this sequence always in practice. An alternative to the adoption process is an "Innovation-Decision Process" in which an individual passes from the knowledge of an innovation to form an attitude to a decision to adopt or reject.

In his book "*Diffusion of Innovations*", Rogers (1995) has proposed different stages in the process.

Knowledge stage: An individual gains knowledge through understanding how the innovation functions. This is "how-to" knowledge usually given by the extension worker at the trial and decision stage. Functioning principles underlying how the innovation works have to be applied by the extension worker.

Persuasion stage: Persuasion makes the individual psychologically involved with the innovation to seek information about the new idea to form a favorable or unfavorable attitude toward the innovation.

Decision stage: A decision is required for the adoption of the new idea to make full use or to reject the innovation.

Implementation stage: Implementation is easy with an individual than an organization where many individuals are involved.

Confirmation stage: The confirmation stage continues after reinforcement of the innovation on the decision to adopt or reject for an indefinite period of time.

Adopter Categories

In a society, it is not expected that all members will adopt new innovations at an equal rate. There are people who can adopt a new innovation immediately, whereas some persons have a suspicious mind or lack of capability for which they may not be willing to adopt any new idea.

The process of adoption of innovations has been studied by many scientists over the years. The adoption index is often calculated to classify the adopter categories. Rogers (2003) describes adoption as a decision to make full use of an innovation and rejection as a decision to adopt an innovation. He defined adopter categories as "the classifications of members of a social system on the basis of innovativeness" which includes innovators, early adopters, early majority, late majority, and laggards. In each adopter category, each individual expresses a similar innovativeness (Table 5.1).

The distribution of innovator categories is based on their innovativeness and can be classified as (i) Innovators (Venturesome), (ii) Early adopters (Respectable), (iii) Early majority (Deliberate), (iv) Late majority (Skeptical), and (v) Laggards (Traditional), according to Rogers and Shoemaker (1971).

- **Innovators:** They are front runners, educated and large farmers willing to adopt innovations first without analyzing the risk. They are usually middle-aged, resourceful and can make immediate decisions. Often they receive information from originating sources without depending only on the extension workers.

Table 5.1: Adopter Category of Farmers

Category of farmers	%
Innovator	2.5
Early adopter	13.5
Early Majority	34.0
Late majority	34.0
Laggards	16.0

- **Early Adopter:** Early adopters are educated, large farmers with high income, influential, good readers, and opinion leaders having farming experiences. They are always willing to test new ideas with caution and prepared to cope with some problems faced in the adoption process.

- **Early Majority:** The early majority may deliberate for some time before completely adopting a new idea. They take relatively longer time than that of the innovators and the early adopters for adoption of the innovations. They are less educated with fewer leadership roles than earlier adopters. They are less active in formal groups than early adopters, but more active than those adopting later. They merely follow the early adopters without any lead.
- **Late Majority:** The late majority comprises people with less education and leadership roles. They wait for some time and adopt new ideas just after the average member of a social system. They adopt the innovations due to economic necessity or under social pressure. Usually, they do not adopt the innovation until most of the social system is done. They can be persuaded of the utility of new ideas, but the pressure of peers is necessary to motivate adoption.
- **Laggards:** They are traditional and orthodox people who come under the last category of adoption of innovations. They have less education and they hardly take part in formal organizations and government programs. They may or may not adopt the innovations. By the time some of them adopt the innovations, the innovators might have switched over to new innovations.

Adoption Curve

The rate of adoption of innovations may vary from individual to individual. Some farmers will be more exposed to new ideas through contact with media messages or with what we call "influential figures." Some may be more ready to try something new because of their better education or financial security. Some may be reluctant to take any risks because of their financial insecurity or their loyalty to tradition. Typically, the adoption process is slow at first. Then, over time, the responses to the recommendations begin to develop a pattern that can be diagrammed as an "adoption curve." Usually, the adoption rate takes an S-shaped curve when cumulative adopters are plotted in a graph over time and frequency basis (Figure 5.1).

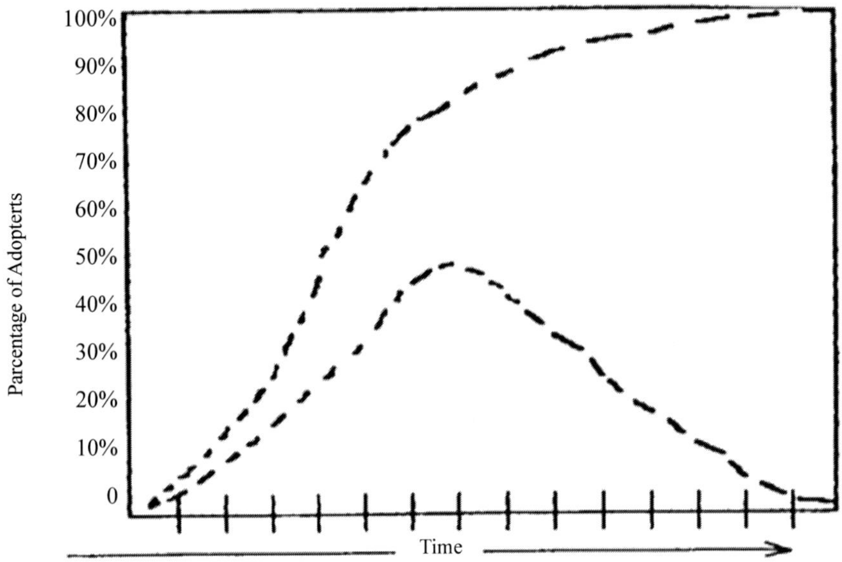

Figure 5.1: The adoption curve of innovations.

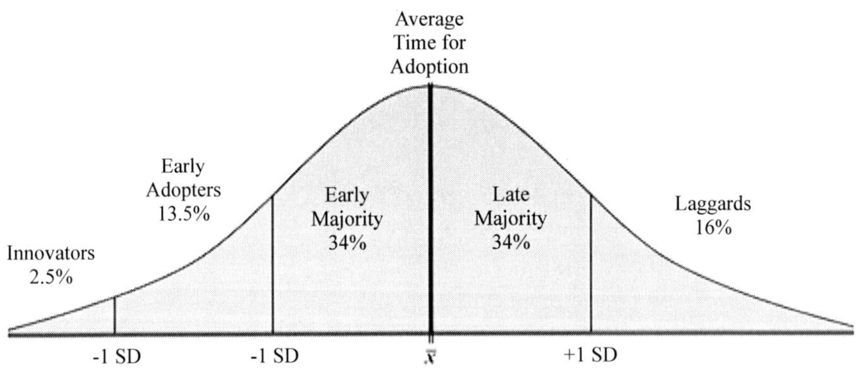

Figure 5.2: Distribution of adopter categories

The distribution may be partitioned into five adopter categories by using the mean (x) and standard deviation (Figure 5.2). The bell-shaped adoption curve (Figure 5.2) indicates the data in terms of the number of individuals adopting each year and the S-shaped curve (Figure 5.1) shows the data on cumulative basis.

The farmers shown at the left of the mean (x) minus two standard deviations are 2.5% of the individuals who are innovators. The next 13.5% between minus one standard deviation and the minus two standard deviations are called early adopters. The next 34% of the adopters between the mean and minus one standard deviation are called the early majority. The late majority lies between the mean and plus one standard deviation. The last 16% located after one standard deviation are the laggards.

Factors Influencing Adoption Process

There are some social, personal, and situational factors that influence the adoption process.

Social Factors

The following are the social factors:

- **Social values:** Differences in communities and their social values influence the adoption process. There are groups who are interested in economic gains whereas in other groups people take it as a prestige in adopting new ideas. There are other groups who value their tradition and customs while adopting new ideas. Hence the extent to which innovations are adopted depends on the values and expectations of the group. The rate of adoption is slow where there is an emphasis on maintaining the traditions and the adoption is quick where emphasis is on economic gain or personal success.
- **Local leadership:** Local leadership is required to influence the adoption. Leadership style of functioning is important to influence the farmers in a community to encourage the adoption of innovations. To influence the neighbors, opinion leaders are to be identified.
- **Social contacts:** Social contact within and outside the community in the diffusion of innovations. Nature and extent of social contact and social distance influence the social contact. Many organizations promote change directly or indirectly. Promotional activities by the organization, broad social orientation, and inter-personnel means can make the social contact more effective.

Personal Factors

Age, educational level, psychological characteristics, and attitudes of individuals influence the adoption process. Comparatively younger people with some education standards and flexible attitudes toward traditional values are seen as ideal factors for better adoption. Elderly people with high traditionalism are not eager to adopt new ideas.

Situational Factors

The nature of adoption, complexity of the practice, cost involvement, net return, compatibility, trialability, and observability are important situational factors for adoption. Greater adoption takes place when the new practice is simple and not complex, involves less cost and fetches more net return, compatible with cultural norms, and has trialability and observability or communicability. Since profitability and simplicity of the innovations are very important, the innovations may be tried in a small area before scaling up.

References

Rogers, Everett M. 1995. Diffusion of Innovations, Fourth edition, New York, Free Press.

Rogers, EM, and Shoemaker, F.F. 1971. Communication of Innovations. New York: The Free Press, New York.

Rogers, E.M. 2003. Diffusion of innovations, 5th ed. New York: Free Press.

Suggested Reading

Anderson, T., Varnhagen, S. and Campbell, K. 1998. Faculty adoption of teaching and learning technologies: Contrasting earlier adopters and mainstream faculty. The Canadian Journal of Higher Education, 28(23):71-78.

Bennett, J., & and Bennett, L. 2003. "A Review of Factors that Influence the Diffusion of Innovation When Structuring a Faculty Training Program." Internet and Higher Education, 6:53-63.

Braak, J.V. 2001. Individual characteristics influencing teachers' class use of computers. Journal of Educational Computing Research, 25(2):141-157.

Dahama OP, Bhatnagar OP. 2005. Education and Communication for Development. Oxford & IBH Publishing Co. Pvt. Ltd., New Delhi.

Dasgupta. 1989. Diffusion Agricultural Innovations in Village India. Wiley Eastern, New Delhi.

Dooley, K.E. 1999. Towards a holistic model for the diffusion of educational technologies: An integrative review of educational innovation studies. Educational Technology & Society 2(4):35-45.

Jalihal KA, Veerabhadraiah V. 2007. Fundamentals of Extension Education and Management in Extension. Concept Publication Company, New Delhi.

McKenzie, J. 2001. How teacher learn technology best. From Now On: The Educational Technology Journal, 10(6). Retrieved March 01, 2005, from http://www.fno.org/mar01/howlearn.html.

North Central Rural Sociology Subcommittee for the Study of Diffusion of Farm Practices. How Farm People Accept New Ideas. Iowa Agricultural Extension Services, Special Report 15, Ames, Iowa, 1955.

Ray GL. 2005. Extension Communication and Management. Kalyani Publishers, Ludhiana.

Reddy AA. 1987. Extension Education. Sree Lakshmi Press, Bapatla.

Spotts, T.H. 1999. Discriminating factors in faculty use of instructional technology in higher education. Educational Technology & Society, 2(4):92-99.

Supe SV. 1987. An Introduction to Extension Education. Oxford & IBH Publishing Co. Pvt. Ltd., New Delhi.

Van den Ban BW, Hawkins BS. 1998.Agricultural Extension. Burnt Mill, Harlow, Essex, England.

6

Reforms in Agricultural Extension in India

Introduction

The history and evolution of extension in India during the Pre-independence and Post-independence periods have been narrated in Chapter 2. The major breakthrough in Agricultural Extension was the introduction of the T&V system with WB support from 1977 to 1997. The system tried to establish a strong linkage between the research and extension. Technical messages received from research stations were transferred to the farmers through an organized extension system at fixed time intervals and obtained feedback from the farmers about the technologies. However, the system was constrained by structural problems at organizational level. The system adopted a classical top-down approach and the recommendations were suitable for irrigated areas bypassing the rain-fed areas. After the withdrawal of support from WB, the T&V system was constrained by the paucity of funds to manage the manpower created, and the unpaid farmers did not continue to work. After withdrawal, it was considered necessary to make the extension more demand-driven, participatory, group extension, and farmer-centric with possible convergence for which extension reform in India was necessary.

Extension Reform in India

There are many countries where extension and advisory services reforms are occurring globally (Swanson and Rajalahti 2010; World Bank 2012). Since agro-ecological conditions and access to markets vary within most countries, decentralized extension system was needed. It was considered that decentralized and demand-driven extension allows farmers to participate in setting the agenda and demanding extension and research priorities. The extension reforms made the system reach the smallholders, resource-poor, and women, and also focused on sustainability. When the donor funds dry up or the pilot projects end, the farmers no longer have access to the extension services (Birner and Anderson 2007). Therefore the public extension needs to be sustainable in extension services delivery.

Toward the late 1990s, two reform initiatives were supported by the WB in the form of the Diversified Agricultural Support Project (DASP) from 1998 to 2004 and the National Agricultural Technology Project (NATP) from 1999 to 2005.

DASP was aimed at increasing the productivity and income of farmers by supporting intensified and diversified agricultural production and farming systems. Based on the experience of DASP by MANAGE, NATP was launched by the Government of India with the financial support of WB from the financial year 1995-96 which was approved in November 1998.

Reform initiatives covered improving the access of women to agricultural extension services. Project (WYTEP) in Karnataka, the 1986–2003 Tamil Nadu Women in Agriculture (TANWA) initiative, the 1998-2003 Training and Extension for Women in Agriculture (TEWA) program in Orissa, and the 1993–2005 Madhya Pradesh Women in Agriculture (MAPWA) project were implemented for this. Next to the gender-focused projects, reform efforts are also directed toward improving the livelihood of disadvantaged tribal groups. Andhra Pradesh Tribal Development Project (APTD) of the International Fund for Agricultural Development (IFAD) was implemented from 1991 to 1998 to improve the income and food security situation and living conditions of tribal groups, taking gender, health, educational, environmental, and marketing and credit issues into account.

Another reform initiative is the ongoing e-Choupal initiative of the Indian Tobacco Company (ITC for improving agricultural extension service provision in terms of outreach (cost) effectiveness, efficiency, transparency, and accountability.

Support to the State Extension Program for Extension Reforms (SSEPER) in 2005, the 2006 National Agricultural Innovation Project (NAIP), and the 2003 National eGovernance Action Plan (NeGAP) are also some of the programs which have not yet been fully evaluated. SSEPER aims to provide decentralized and demand-driven extension services through the active participation of farmers, extension specialists, civil societies, and agricultural science centers in 28 states and two union territories. NAIP emphasizes agricultural research and development (R&D) rather than extension.

National Agricultural Technology Project (NATP)

The broad objectives of the NATP initiative were (1) to improve the efficiency of the organization and management systems of ICAR, (2) to strengthen the effectiveness of research programs and the capacity of scientists to respond to the technological needs of farmers, and (3) to increase the effectiveness and financial sustainability of the technology dissemination system with greater accountability to and participation by farming communities. The NATP basically is a technology project which focuses on research and extension and on an integrated system of extension delivery.

Both the DASP and NATP projects provided supply-side and demand-side service delivery. On the demand side increasing agricultural production and growth and rural development were emphasized through a broad-based, bottom-up, and participatory approach involving community groups. On the supply side public expenditure reform, reforming planning and decision-making processes, and training and capacity building of stakeholders were taken.

The Agricultural Technology Management Agency (ATMA) model was particular to the DASP initiative in Uttar Pradesh. Uttaranchal did not undertake the ATMA approach but implemented Agricultural Diversification Management Societies, which were established at each of the project districts to coordinate, plan, and implement the project through officers of line departments and farmer federations.

Under the NATP initiative, there was decentralization in both extension and research. The components under NATP were (i) Agro-ecosystem Research (AER), (ii) Innovations in Technology Dissemination (ITD), (iii) Organization and Management System (O&M), and (iv) Information System Development (ISD).

The Innovations in Technology Dissemination (ITD) component of NATP emphasized on technology dissemination system at the national, state, district, block, and village levels. The Technology Dissemination Management Committee (TDMC) and the Technology Dissemination Unit (TDU) were strengthened under this component for monitoring, supervision, and evaluation roles for technology dissemination at the national level.

State-level Organization

The Interdepartmental Working Group (IDWP) at the state level harmonized the project activities of the line departments. In addition to interaction with the TDMC, the IDWP also monitors the progress of district-level programs implemented by ATMAs. Technology dissemination at the district, block, and village levels is managed by the ATMA.

Under Rashtriya Krushi Vikas Yojana (RKVY), a State Level Sanctioning Committee (SLSC) is constituted to approve the State Extension Work Plan (SEWP). The State Nodal Cell (SNC) will ensure timely receipt of District Agriculture Action Plans (DAAPs) and formulate the State Extension Work Plan (SEWP) with the help of the State Farmer Advisory Committee (SFAC) and get the approval of the SLSC. It is the responsibility of the

SNC to communicate the approval of the work plan of ATMAs and the State Agricultural Management & Training Institute (SAMETI). The SAMETI prepares the training calendar accordingly.

District Level ATMA

At the district level Agricultural Technology Management Agency (ATMA) was introduced as a pilot (1998-2003) in 28 districts (DAC 2005) in seven states which was scaled up across 251 rural districts in 2005 and throughout the country in 2007 (Reddy and Swanson 2006).

The ATMA is an autonomous institution formed at the district level for the management of technology dissemination, decentralized planning and implementation, and promotion of inter-departmental coordination at the district, block, and village levels involving line departments, NGOs, and farmer groups. ATMA exercises full discretion over its budget and flexibility in changing technological and environmental requirements.

ATMA is assisted by a Governing Board (GB) and the ATMA Management Committee (AMC). The Governing Board recognizes the programs and procedures for district-level research and extension activities, assesses the progress and functioning of the ATMA, and approves the Strategic Research and Extension Plan (SREP) prepared by ATMA. The AMC oversees the implementation of research and extension programs and SREP.

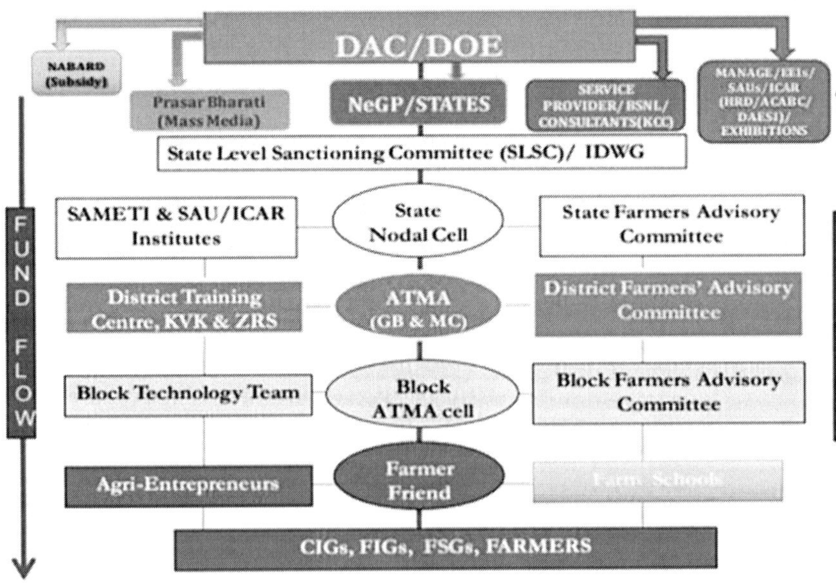

Figure 6.1: Organizational structure of ATMA.

Block Level

A Farm Information and Advisory Centre (FIAC) created at the block level implements all programs as per SREP with the help of the Farmer Advisory Committee (FAC).

Block Technology Team (BTT) is the advisor of the Farmer Advisory Committee (FAC). The FAC interacts with all key stakeholders by forming commodity-oriented farmers' and women's interest groups at the block and village levels.

Village/Panchayat Level

Group extension is managed at the GP/Village level by several farmers' groups like Commodity Interest Groups (CIGs), Farmer Interest Groups (FIGs), and Food Security Groups (FSGs). Farmer Producer Organizations (FPOs) are also formed to provide backward and forward linkage. A Farmer friend (Krushak Sathi) is posted for every two villages to serve as a link between the farmers and the extension system with regard to the dissemination of technology. Supporting extension is available from Agri-entrepreneurs, Agri-clinics, and input dealers.

Revised guidelines for ATMA were issued by the Department of Agriculture and Cooperation (2010) to strengthen ATMAs by making the extension system farmer-driven and farmer-accountable. Dissemination of technology to farmers was made through ATMA at the district level to operationalize the extension reforms in a participatory mode under the ITD component

Involvement of Private Sector

Effective involvement of public, private, and third-sector service providers improves decentralization in research and extension. The supply-side governance mechanism emphasizes the role of the private sector as an instrument to supplement the extension and research efforts of the public sector and to improve the efficiency and effectiveness of service provision (Reddy and Swanson 2006). They can play a critical role in providing demand-driven support services and encouraging the introduction of improved inputs (such as fertilizer and seed varieties), and strengthening marketing and postharvest activities and processing. The NATP also promoted private-sector participation in policy formulation through formal consultative mechanisms (World Bank 1998a). The involvement of NGOs is important because their work centers on clearly defined and specific objectives (Shekara 2001; Shankar 2001; and Nataraju *et al.* (2001).

Public-public Partnership

The public-public partnership between the state Department of Agriculture and the State Agricultural University operated in a District Agricultural Advisory Technology Centre in all districts of AP. These centers are required to refine the technology, make diagnostic visits, and organize field programs in collaboration with the Department of Agriculture and allied departments to improve the dissemination of technology, strengthen research capacities, and foster human capital accumulation at ICAR institutes. Some States conduct adaptive research, training, and consultancy at the district level by having a multidisciplinary extension team in each district, again in collaboration with the state Department of Agriculture.

Training of Extension Workers

The training of extension workers is carried out by the SAUs, the State Institute for Rural Development, SAMETI, and the National Institute of Agriculture Extension Management (MANAGE). These institutions offered professional skill training to extension personnel across all line departments and Krishi Vigyan Kendras, forwarded technical information, and supported the technical training of district-level subject matter specialists and of the instructors of SAMETI.

Under the NATP, a critical source of human capital development was the National Institute of Agriculture Extension Management (MANAGE). MANAGE's main task was to enable the project-participating states to develop a decentralized, location-specific, farmer-centered sustainable technology dissemination system. NATP strengthened the role of MANAGE as a provider of training in extension management, strategic planning, and participatory rural appraisal to senior managers and trainers at the national and state levels. MANAGE provided training to extension workers in the first and second phases. In the third and fourth phases, Agricultural Management and Extension Training Institutes (SAMETIs) were taking these activities. In addition to the SAMETIs, extension workers also received training from the ATMAs.

ICT

The private sectors rely much on ICT for efficient communication means. e-Choupal initiative of the Indian Tobacco Company (ITC) was launched in 2000. It now operates in 31,000 villages in six Indian states through 5,200 kiosks, extending to 3.5 million farmers. The e-Choupal initiative is kiosks in rural farming villages that are equipped with computers with internet access via phone lines or by a Very Small Aperture Terminal (VSAT) connection. As villages often face power supply shortages, each e-Choupal is equipped with a battery-based uninterrupted power supply backup, which provides sufficient energy to run the system twice a day.

The NATP initiative, for example, actively encouraged the use of ICT facilities in order to promote the organizational and managerial efficiency of the ICAR research system and in order to strengthen the link between the research and farming system and the research and extension system.

Activities Under Support to States for Extension Reform

The government of India has indicated the cafeteria of activities under Extension Reform at the state, district, and block levels. The activities are included in four components, such as (i) Specialist and Functionary Support at various levels, (ii) Cafeteria of Activities, (iii) Innovative Technology Dissemination Activities, and (iv) Increasing Coverage of Activities, Inclusion of New Activities, and Infrastructure. The States are allowed to choose locally suitable activities from the Cafeteria in keeping with the priorities of the states.

State level: Support is given for infrastructure development, human resource development, and other activities like organization of Agricultural Exhibitions, and Fairs, providing incentives for good extension activities, prizes to farmers, M&E of Schemes on extension, and Induction and Refresher training of ATMA staff.

District/block level: Specific activities have been suggested under (i) Farmer Oriented Activities; (ii) Farm Information Dissemination; (iii) Research-Extension-Farmer (R-E-F) linkages; and (iv) Innovative Technology Dissemination activities. These activities are included in the Strategic Research and Extension Plan which is prepared by ATMA. The activities emanating from the Strategic Plan and ITD interventions are implemented by the ATMA. Involvement of Agri-clinics and Agri-business Centers, trained Input Dealers, REF linkage, and setting up Community Radio Stations are included with the cafeteria of activities.

The success of extension system reforms depends on an effective linkage with research. The reform measure from ICAR relating to the implementation of the extension reform was the issue of a set of directives jointly with the Department of Agriculture and Cooperation (DAC-DARE, 2011; ICAR, 2011).

Supply-based and Demand-based Extension

The current extension system needs to operate beyond the transfer of technology. Activities like value chains will require technical expertise that goes beyond the capacity of the current extension functionaries. The extension staffs require training for such innovations, and hiring experts at the district and block levels for such services with the involvement of NGOs. This partnership with the private sector is required in agricultural extension services.

Market-led Extension

Market-led extension requires (i) assessing the potential markets for different high-value crops or products that can be successfully produced in different blocks within the district, (ii) identifying products that can be easily marketed, (iii) identifying more promising crops or products that are well suited to the district and (iv) diversify into a variety of different high-value crops/products that are suitable for different Farmer Interest Groups (FIGs) or Women Interest Groups (WIGs) within the district. The first set of products, markets, and supply chains to be examined is the "success stories" that are identified to determine whether each success story provides scope for marketing. The markets for higher-value products are expanding rapidly in India as the urban middle class has more disposable income and desires new products. Identification of the suitable markets, competitive advantage, and the producing farmers is the primary job of ATMA. The farmers can be organized into FIGs and WIGs and linking them with FPOs. Required training and exposure visits could be arranged for these groups. The local KVKs can facilitate these groups in training, exposure visits, and supply of critical inputs.

Strategic Research & Extension Planning (SREP)

Under NATP, ATMA is mandated to prepare a SREP for the district. SREP identifies the research and extension gaps in agriculture and allied sectors in the district and prepares a holistic plan for the future growth of the organization.

The SREP indicates organization's mission and goals and programs and services required to achieve the goal and how to mobilize and utilize resources, manpower, funds, expertise, and facilities in the future.

Rationale for SREP

SREP is a future roadmap of the organization which lays down the futuristic growth of the organization. Based on the needs of the stakeholders and problems, it gives direction for resource utilization, addresses the environmental uncertainties, and provides important decisions and actions for the development of agriculture and allied sectors. Following a bottom-up approach in planning, it analyses the present scenario of the district in terms of potential, opportunities and problems in agriculture and allied sectors under different farming situations. The situation analysis facilitates a long-term vision and strategic planning in a participatory manner. The annual action plan is prepared for each block based on the SREP.

The scope of the SREP is as follows:

- To be strategically alert.
- To be future-oriented.
- To bring convergence.
- To generate large resources, and
- To be effective in technology, extension, and market support systems.

The Block Action Plan (BAP) is formed based on the SREP on an annual basis. The District Level Action Plan (DAAP) is a subset of Comprehensive District Agriculture Plans (CDAP) prepared for the District under (Rashtriya Krishi Vikas Yojana (RKVY). The DAAPs will be consolidated in the form of a State Extension Work Plan (SEWP) as a part of the State Agriculture Plan (SAP) (Figure 6.2).

Contents of SREP

The SREP has both diagnostic and strategic sections. In the diagnostic section information and analysis are included while the strategic section includes strategies and action programs.

(i) Diagnostic section

In the diagnostic section, basic information about the district is collected which is then systematically analyzed for developing strategies. The district information is collected from secondary data while the information for representative villages under each agro-ecological Zone is collected from primary sources using Participatory Rural Appraisal tools and techniques.

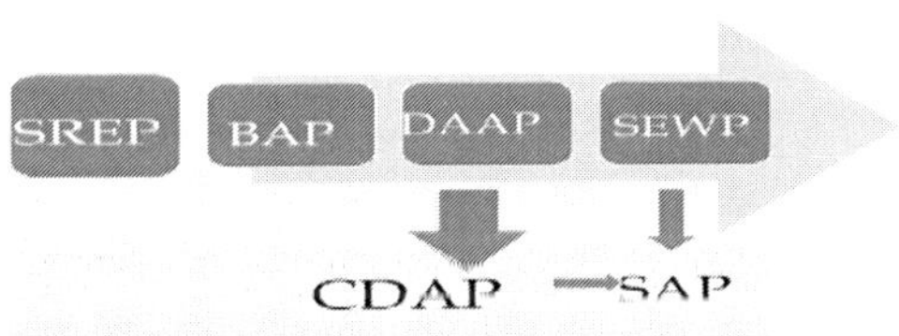

Figure 6.2: State Agricultural Plan.

District Information: Under district information, general features of the district, agro-ecological information, demographic data, description of land-based systems, infrastructure facilities, soil data, water resources, vegetation, the area under irrigation and rain-fed farming, ongoing development programs, marketing infrastructure, agro-processing facilities, agricultural credit, input supply, farmers' organizations, private sector organizations and NGOs and ICT are included.

Information from Representative Villages under Each AES: As per guidelines issued by MANAGE the information on geographical area, population structure, the area under rain-fed farming and irrigation, households, land and soil, farming system, marketing, agro-processing, storage, input service, farmers' organizations, and private sector organizations are collected.

Based on the data collected from secondary and primary sources analysis of the farming system is made.

The analysis of the existing farming systems, the influence of resource situation on the existing farming system, and SWOT analysis of different farming systems are done by adopting different techniques.

(ii) The Strategy section

After analyzing the existing farming systems and farming situation for each AES (Figure 6.3), the strategies and activities are formulated to address the research and extension gaps. Research activities would be prioritized with the help of KVKs. Both Extension and Research strategies are formulated after this.

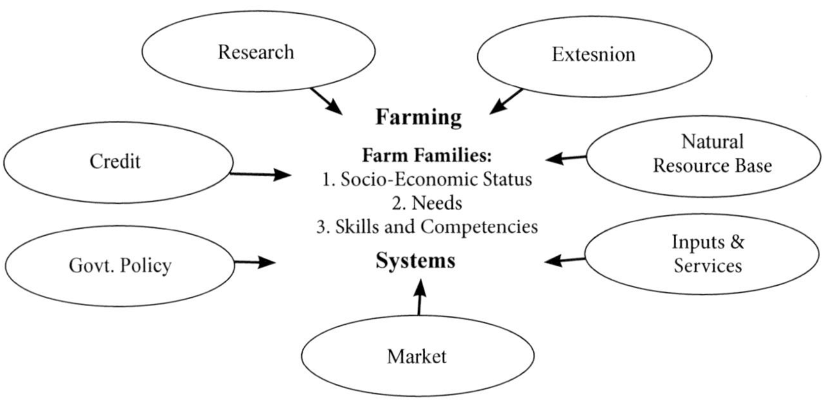

Figure 6.3: Farming system analysis.

Extension Strategies should cover diversification and intensification of existing farming systems, sustainability of extension, human resource development, the convergence of various ongoing schemes at public, private, and NGO levels, promotion of Market-led extension, Public-Private Partnership, development of ICT, Gender mainstreaming, etc.

Research strategies would cover the re-synthesis of the technological package as per the farming situation of each commodity, farmer participatory on-farm research on those problems that cannot be simulated in research farm, and on-station research on other problems.

Activities

As per different strategies, various activities will be planned. For example, in agriculture and horticulture production systems the activities could be expansion of area, substitution, improvement in rain-fed farming, diversification, increasing cropping intensity under irrigation, input use and adoption of technology for yield improvement, post-harvest management, etc. could be indicated. Various activities under each strategy have to be clearly mentioned for implementation. The program thrusts in SREP of each district will cover farming system innovations, farmer organizations, addressing technical gaps, and management of natural resources.

Process of Developing SREP

Preparation of SREP involves eight steps as outlined under:

Step 1: Mobilization of the resource team and their orientation

Step 2: Identification of agro-ecological-situations by constitution of AES teams and logistic arrangements

Step 3: Training of AES teams

Step 4: Participatory Field data collection

Step 5: Rechecking and analyzing data and information

Step 6: Developing strategies for research and extension

Step 7: Developing Activity Schedules

Step 8: Approval SREP

MANAGE has developed a checklist for all these steps that may be used while preparing the SREP that needs to be followed while preparing the SREP of the district.

Operationalization of SREP

After approval of ATMA, the SREP becomes a basic document for implementation in the district.

Action planning: The action planning gives an outline of functional attributes in terms of what, when, where, who, (for) whom, and how long with financial modalities. The copies of the SREP will be distributed to all concerned for implementation.

- Block Technology Team (BTT) members shall identify activities of SREP that are relevant to (AESs) their block
- SREP acts as a guiding force for identifying the activities. Both the BTT and FAC shall attend to the following while preparing the Block Action Plan.

- Extension activities identified in SREP;
- Group demands as identified by FAC members;
- On-going schemes for dovetailing;
- Success stories identified for replication.

The above four dimensions will be considered on an annual/seasonal basis in the joint meeting of BTT and FAC for the preparation of BAP.

- The Farmer Advisory Committee (FAC) will approve the BAP as per funds availability and send it to ATMA.
- The BAPs of the district will be approved by ATMA in AMC.
- ATMA will prepare its ATMA level action plan based on the needs and priorities cutting across blocks boundaries and issues emanating out of SREP referring to the cafeteria of activities.
- ATMA will prepare the District Action Plan which would be approved by the GB. GB will ensure that there shall be equitable fund allocation across the blocks.
- All DAPs are for preparation of SEWA at the State Level with the assistance of SAMETI.

Fund flow mechanism: As per the primary document circulated by GOI under the 10th Plan Scheme "Support to State Extension Programs for Extension Reforms," based on the SEWP the funds would be placed with an autonomous institute like ATMA and SAMETI. The funds received by ATMA will be passed to FACs for the execution of programs at the block level. Necessary records will be maintained at the ATMA and FIAC levels.

Execution of Extension Programs: Once the funds are received by the Convenor, BTT, the field program needs to be executed. Extension Programs like Awareness Camps, Exposure Visits, Demonstrations, and Training, etc.

Execution of Research Programs: Researchable issues identified in SREP shall be referred to the Zonal Agricultural Research Extension Council (ZAREC) of ZRS or the Scientific Advisory Committee (SAC) of KVK. The issues that have been emerged during SREP would be categorized into (i) long-term researchable issues and (ii) short-term researchable issues. The long-term issues shall be communicated to SAU, and ICAR institutions for redressal.

Monitoring of field activities

- The block-level and village-level activities would be monitored on a monthly basis at the FIAC level. The progress made shall be reviewed during the first week of every month in the joint meeting of BTT and FAC members. The Convener of BTT shall compile the progress reports.

- The review of progress should be undertaken by AMC at the ATMA level during the second week of every month after receipt of progress reports from the blocks. The quarterly review will be made by the GB.
- The district Collector will review the programs on a quarterly basis in a joint workshop GB of ATMA & FAC.
- State-level monitoring will be undertaken by IDWG through reporting and workshops.

Strengthening SREP with agricultural marketing plan

Convergence amongst agriculture and allied departments is required in the marketing of agricultural produce. The marketing component is incorporated into the SREPs by bringing together the concerned agencies on a single platform. The following steps are to be taken for inclusion of the marketing plan in SREP.

- Interaction between agriculture & marketing department
- Core team formation
- Orientation of core team
- Data collection for the marketing component
- Analysis & consolidation of the data.

Modified Extension Reform

Modified Extension Reform Schemes were initiated in 2010 by amalgamating 17 schemes of the 11th Plan period. The new mission has been named as National Mission on Agricultural Extension and Technology (NMAET) with four submissions. The common programs in these sub-missions are Extension & Technology. The mission aims at restructuring and strengthening agricultural extension for the transfer of appropriate technology to farmers.

Sub-Mission on Agricultural Extension (SMAE) focuses on awareness creation and enhanced use of appropriate technologies in agriculture & allied sectors including capacity building of personnel under the Agri-Clinics and Agri-Business Centres Scheme (ACABC) and Diploma in Agriculture Extension Services for Input Dealers (DAESI).

Sub-Mission on Seed and Planting Material (SMSP) covers seeds and planting materials provides required infrastructure support and envisages strengthening of Protection of Plant Varieties and Farmers" Rights Authority (PPV&FRA) to protect the rights of farmers and plant breeders in evolving new varieties.

Sub-Mission on Agricultural Mechanization (SMAM) emphasizes the promotion of farm mechanization to cater to the needs of small and marginal farmers by establishing custom hiring centers in selected villages and disbursing subsidies for the procurement of machines & equipment.

Sub-Mission on Plant Protection (SMPP) focuses on strengthening Integrated Pest Management (IPM) and introduction of modern plant protection approaches. Residue analysis of pesticides is also included in the sub-mission. The component of the National Institute of Plant Health Management (NIPHM) promoted environmentally sustainable plant health management practices in diverse and changing agro-climatic conditions.

References

Birner, R., and J. Anderson, J. 2007. How to make agricultural extension demand-driven? The case of India's agricultural extension policy. Discussion Paper 00729. Washington, D.C.: International Food Policy Research Institute.

DAC-DARE. 2011. Convergence between Research and Extension. New Delhi: Department of Agriculture and Cooperation and Department of Agricultural Research and Education. Ministry of Agriculture, Government of India, New Delhi.

DAC 2014. ATMA Guidelines, 2014 under NMAET.

DAC Department of Agriculture and Cooperation 2010. Guidelines for Modified Support to State Extension Programmes for Research and Extension Reforms Scheme, 2010, Department of Agriculture and Cooperation, Ministry of Agriculture, Government of India, June.

MANAGE – Strategic Research Extension Plan (SREP) Guidelines (2005) SREP documents of ATMA districts.

Reddy, M. N., and B. Swanson.2006. "Strategy for up-scaling the ATMA model in India." Annual Conference Proceedings of the Association for International Agricultural and Extension Education 22: 561–569.

Shankar, K. R. 2001. "Private extension in India and a SWOT analysis." In Private extension in India: Myths, realities, apprehensions and approaches, edited by. P. C. Shekara. Hyderabad: National Institute of Agricultural Extension Management (MANAGE).

Shekara, C. 2001. "Private Extension: Indian Way." In Private extension: Indian Experiences, edited. by P. C. Shekara. Hyderabad: National Institute of Agricultural Extension Management (MANAGE).

Swanson, B. and Rajalahti, R. 2010. Strengthening Agricultural Extension and Advisory Systems: Procedures for Assessing, Transforming, and Evaluating Extension Systems, Agricultural and Rural Development, Discussion Paper 45,. Washington, DC: The World Bank, Washington DC.

World Bank. 2012. Agricultural Innovations Systems – An Investment Source Book., Washington, DC: The World Bank, Washington D.C. 658.

Suggested Reading

Ban, A W Vanden., 2006: Changing roles of Agricultural Extension in Asian Nations: Delhi: B.R. Publishing Corporation.

Desai, G R. 2004. Reforms in Extension: Experiences from ITD component of NATP in Orissa Bhubaneswar: Institute on Management of Agricultural Extension, 2004.

IIM (Indian Institute of Management) 2004. Impact Assessment Report on the Innovations in Technology Dissemination (ITD) Component of the National Agricultural Technology Project, Indian Institute of Management, Lucknow.

Nataraju, M. S., M. T. Lakshmi Narayan, and K. H. Nagaraj. 2001. Role of Non-governmental Organizations in the New Millennium – Case studies. In Private extension: Indian experiences, ed.

Sharma, S. R. 2002. New Perspectives on Extension Education. Jaipur: Book Enclave, 2002.

Shingi, P M. 2004. State of the Indian Farmer: Agricultural Extension New Delhi: Academic Foundation.

Singh, J. P., B. E. Swanson, and K. M. Singh. 2006. Developing a decentralized, market-driven extension system in India: The ATMA model.

Suresh C. Babua et. al., 2013. The State of Agricultural Extension Reforms in India: Strategic Priorities and Policy Options; Agricultural Economics Research Review 26 (2):159-172.

Van den Ban, A W. Agricultural Extension - 2nd. New Delhi: CBS Publishers & Distributors, 1998.

7

Participatory Extension Management

Approaches in Extension

Agricultural extension services have gone through various stages over the years, from the individual farm visits approach to the T&V approach. All these approaches are characterized to a greater or lesser extent, by a "top-down" planning. Extension approaches mostly determine the direction and nature of the system in terms of structure, leadership, program, methods and techniques, resources, and linkages. Extension approach is a style of action within a system which embodies the philosophy of the system. Although there are many approaches to agricultural extension, there is certain commonality in them. All approaches function through non-formal education and all have agricultural content for the development of rural people.

Ministry- based General Extension

Ministry of Agriculture was entrusted with agricultural extension in a top-down manner for planning for the adoption of technologies. An increase in national production is taken as the measure of success.

All persons engaged in agriculture are the clients of this public extension who are served by extension staff. Program formulation and implementation are guided by public interest. Ministry-based extension faces difficulty in serving the majority of potential clientele and financial constraints that force it to reduce the staff. Whereas extension demands location-specific technologies and flexible decisions, the university extension is mostly top-down in nature. As the approach lacks a two-way approach the messages are not location specific. Production alternatives are absent in this approach.

The Training and Visit Approach (T&V)

This is a technology transfer approach (TOT) where research findings are transferred to farmers through the extension for the solution of farmers' problems. Research is carried out at research stations followed by field trials in farmer's fields. Once a possible technical solution is identified and proven, it is developed into a recommendation for the farmers. Extension workers usually disseminate this 'new knowledge' to the farmers to solve their problems with the expectation that the newly recommended practice will diffuse amongst the farmers. In this system, the farmers are found as passive receivers of technologies, and the technologies are decided by the specialists.

The T&V system was developed in Turkey in 1967, successfully used in countries like India in the 1970s, and subsequently used as a blueprint in other countries. This approach was functioning in more than 40 developing countries. It is a system, which emphasizes simplicity in both objectives and operations. This approach uses a planned schedule of visits to farmers by trained extension workers and subject matter specialists. The system establishes a close link between research, extension, and farmers. Technology transfer is the main responsibility of the extension workers. This is a top-down approach and a two-way channel is somehow missing. The contact farmers work as link persons between the extension workers and the farmers. It is built on a combination of individual and group approaches. The technical capability of the extension service was enhanced in the system due to systematic training and regular farm visits.

A review of the T&V system found that it was not flexible and the extension workers covered many topics without addressing the major concerns of farmers. Many farmers were excluded, either by selective contact or by the inappropriateness of the packages. The contact farmers selected by the extension workers were found to be men with more formal education and larger farms than non-contact farmers. The resource constraints faced by many of the non-contact farmers impeded the adoption of many new technologies. Additionally, extension workers also felt that they were being asked by their supervisors to achieve targets rather than being evaluated on whether they had transferred any relevant information to farmers. With the pace of time, the un-paid contact farmers did not take interest being associated with the TOT model. After the withdrawal of external support, the system was constrained by a shortage of staff and funds. However, there are situations where the T&V system has been effective due to the availability of inputs and marketing.

The Integrated (Project) Approach

Integrated project approaches aim at alleviating mass poverty in rural areas with an emphasis on self-help. The evaluation shows that large numbers of poor farmers were not reached by project activities, nor were positive effects consolidated on a sustainable basis. Management deficiencies, ambitious goals and complexities in the program are found in this system. The key concept is the availability of locally adapted solutions established on a common basis which requires not only participatory technology identification, test, and dissemination, but also an active role by the agency in mediating between different institutions involved and their interests. New efforts would be required to specify and operationalize extension objectives and concepts (sustainability, participation, gender-specific target-group approach, and poverty alleviation).

University-based Extension

This approach is adopted by the Universities of educational institutions with technical knowledge and research abilities. The implementation is through non-formal education in groups or individuals by the college or university. Through their extension specialists, they transfer technology to the farmers through extension workers and receive feedback. The extent of participation and attendance of farmers in agricultural extension activities are the measures of success. The advantage of this approach is the relationship developed between the research scientists and field extension personnel.

New findings and feedback are presented, evaluated, and published as a "Package of Practices" to be used by all extension staff for the next season (Nagel, 1980). In University extension, Krishi Vigyan Kendras (KVKs) have been established by Universities, ICAR Institutes, and even by some NGOs for organizing training, technology transfer, conducting frontline demonstrations, and supply of critical inputs. However, the system varies from state to state. The Punjab Agricultural University (PAU) has its own multidisciplinary extension team in each district, engaged in adaptive research, training, and consultancy. Backed up by extension specialists on campus, they are transmitters and receivers of experiences from researchers, farmers, and state extension workers. Regular workshops are held which unite university and department staff from research and extension together with outstanding farmers. New findings and feedback are presented, evaluated, and published as a "Package of Practices" to be used by all extension staff for the next season (Nagel 1980).

Commodity-based Extension

The commodity-based extension is implemented by government, parastatals, or private firms in which all aspects of producing and marketing one or two crops are vertically integrated. All functions of production including extension, research, input supply, marketing, etc. are under a single administration. Although the extension is centralized, is commodity oriented. Trained specialists with scientific tools recommend technologies for the greater interest of the commodity production organization. However, the technologies are narrow-based. They are largely defined from the perspective of the commodity organization.

Farming Systems/Participatory Methods

Researchers working around the world noticed the unique problems of the small-scale farmer livelihood system and developed strategies to solve these problems. Hence a system was implemented in a holistic and system approach implemented through a partnership of research and extension personnel. The program is jointly controlled by local farm families, extension officers, and researchers. The system establishes a strong linkage between extension and research, and the commitment of the farmers towards technologies. This system emphasized on smallholder farmers and their livelihood systems as they were excluded from the traditional extension. However, the small-scale farmers living in risk-prone and complex environments were unable to take advantage of the technologies developed in research stations. This approach emphasizes strong on-farm research components where farmers, extension workers, and researchers work together as a team.

Extension as a Commercial Service

Commercial extension provides a specialized consultancy service demanded by an agricultural producer. The objective of this extension is the optimal utilization of purchased inputs or contracted expertise. As the system is costly, it does not serve public interests.

Client-based and Client-controlled Extension

This is a decentralized system of extension that serves the needs of specific target groups better. The extension workers have close contact with their clients and intimate knowledge of their life situations which is essential for the planning of problem-oriented extension activities. After the withdrawal of organizations, local leaders take over the extension. The system works for the creation of awareness, empowerment, participation, and self-help of the clients. The system cannot serve a large number of farmers.

Target Group Approach

Execution of extension work in this system specifically aims at one or several categories of farmers, rather than at the farming population as a whole. The more intensive techniques were designed for larger-scale and more market-oriented farmers. After the minimum resources needed and the technology package is developed, the right farmers for the dissemination of information are selected.

Cost-sharing Approach

In this approach, a part of the cost of extension is shared by the local people. The approach is based on the assumption that cost-sharing with local people will promote a program that is more likely to meet local situations and where extension workers are more accountable to people. The willingness and ability of farmers to share the cost is a measure of success. It invites problems when local farmers are pressurized for invest in unproven enterprises.

Participatory Agricultural Extension Approach

The approach assumes that the active participation of farmers is necessary for group learning and group action. The extension activities are group meetings, demonstrations, individual and group travel, and local sharing of appropriate technologies with a focus on the expressed needs of farmers for improving production and quality of rural life. It is a decentralized and flexible approach and the success is measured by a number of farmers actively participating. The indigenous knowledge is combined with scientific knowledge. This approach focuses on the felt needs of farmers' groups and its goal is increased production and improved quality of rural life. Implementation is often decentralized and flexible. The participatory agricultural extension approach costs less, fits needs well, and is more efficient. Local people can evaluate their own programs. However, this increases the workload of the extension workers, who need to socialize themselves.

Project based approach: This approach operates in a project mode on a particular location for a specific period with outside resources. The approach aims to demonstrate the practices which could be sustainable after the project period. The project is centrally controlled with outside funding. The benefits are measured as short-term success.

Agriculture Extension System by Non-public Players

In addition to public extension services, various non-public players are also involved in agricultural extension.

Input dealers: Input dealers who are marketing seeds, pesticides, fertilizers, and farm machinery are engaged in agricultural extension. A study estimates that 13% of farmers get their information from input dealers in the country (Singh et al. 2016). However, their aim is the promotion of their own materials. Recognizing the importance of this category of extension support to farmers' training are given by MANAGE and other recognized institutes on input use efficiently and to certify them as qualified input dealers.

Agribusiness companies: Agribusiness and input manufacturing companies like ITC, Nuziveedu Seeds, IFFCO, KRIBCO, Pepsico, Heritage, etc. are also undertaking extension activities in support of their product brand and seek to help the farmers to realize higher production

NGOs: Many NGOs like PRADAN, BAIF, etc. are actively involved in the development of rural areas and naturally orient themselves toward land-based livelihoods. Hence, they accommodate an essential component of extension in their intervention (Ashok Gulati et al. 2018).

Organized user groups: Some of the most successful farmer organizations in Maharashtra are MAHAGRAPE, and the United Planters Association of Southern India (UPASI) in providing research and extension services to farmers. Now Government of India is promoting Farmers' Producers' Organizations (FPOs) to strengthen extension activities.

Extension through media and ICT: An increasing number of ICT tools such as mobile phones, internet, TV, KCC, etc. has shown tremendous potential to disseminate information to farmers.

Agricultural Extension Models

A model is a schematic description of a system that may be used for further study of its characteristics. In extension, a number of models have been used to enhance the effectiveness of extension service delivery.

Technology Transfer Model

Extension organizations usually focus on technology transfer and human resource development. The conventional extension has used top-down, TOT methods for extending new technologies. In this approach, technologies are generated at research stations and diffused to farmers using the extension service (Put 1998). The TOT model is based on the assumption that the transfer of technology and knowledge from scientists to farmers will ensure agricultural development. The approach makes the researcher's task to identify, analyze, and solve farmers' technical problems while the solutions are developed in research stations and findings are transferred to farmers as messages by extension works. Hence the extension workers and the farmers have been passive recipients of technologies in a top-down manner without a feedback system. Rogers himself moves away from linear technology transfer with the convergent model in the latest version of his theory on the diffusion of innovations (Rogers 1995).

The Public Extension Model

In the public extension model, public interest was considered for goal setting, program formulation, and the implementation of fieldwork. Public interest aims at serving farmers and the urban population for agricultural production, and promoting cash crops for export, and serving the needs of specific groups. According to the ministry, extension is not only an educational activity and non-educational activities may be included in this. The ministry-based extension is unable to reach a majority of its potential farmers for economic, socio-psychological, and technical reasons. Financial constraints have compelled for reduction of staff at the field level. Very often the location-specific solution to a farmer's problem is not available as it has been a research concern or the solution has not reached the field.

Commodity Extension Model

Commodity-based extension is adopted by governments, parastatals, or private firms. In this model, producing and marketing a particular crop are vertically integrated, starting from research, advice, and material support for marketing and export. The focus on only one or two crops facilitates the training of extension workers. The rigidity of the system fails to incorporate farmers' needs. Extension workers are regarded as successful when they convince the farmers to produce "what and how the organization wants." Many governments have used the approach to excessively extract revenue by dictating low farm-gate prices.

T&V Model

The T&V initially became strikingly original and promising. Rather than trying to reach all farmers directly, the system concentrates on contact farmers as link between extension workers and other farmers. Although regular training and visits were ensured and unified extension service was in place, a two-step flow of information from the extension worker to the contact farmer and from there to other farmers has frequently failed. Communication skills, leadership and organizational capacities are neglected and the system has been a top–down approach without any mechanism for critical feedback based on self-evaluation. This resulted rigidity rather than any flexibility to fit into local situation. The standardized messages were not relevant to local conditions. Applied less rigidly and combined with the tools of human resource development as well as with the concept of participation, these elements may constitute a valuable base for reforming extension organizations, large or small (Nagel 1992).

Farmer Field School (FFS) Model

This model started in the Philippines and Indonesia in the late 1980s on Integrated Pest Management (IPM) practices for rice. The approach uses participatory methods 'to help farmers develop their analytical skills, critical thinking, and creativity, and help them learn to make better decisions' (Kenmore 2002). The method intends to train adult farmers in an informal setting within their own environment as a field school without walls. The field schools are oriented to providing the knowledge and management skills in a participatory manner to empower farmers to be their own technical experts on major aspects of crop and livestock production. FFS is a group extension process based on non-formal education methods, focusing on field observations, season long research studies and hands on activities. During the process, it provides a learning environment and attempts to build the capacity of the group.

Innovative Linkage Models

Most recently, extension is working for linking research with community group needs and helping to facilitate appropriate technology development. Participatory approaches are upheld and linked to wider concerns for strengthening the rural poor's capacity to manage their resources more sustainably and also to articulate their demands to advocate and negotiate for programs that fulfil their needs. Improvement in institutional relations for the technology transfer process is made by collaborative relations between groups and farmers, and new forms of interaction and learning.

Group Formation for Participatory Research and Extension

Social capital in the form of groups is formed for safety nets to cope with risks and for mutual assistance. People are formed into homogeneous groups according to gender, wealth, and age.

Farmer groups: Farmer groups are formed for collective action in areas of agricultural operations, accessing credit and information, etc. Groups allow farmers to obtain new technologies, benefit from economies of scale, enter into stable relationships with suppliers, and set rules for natural resource management (Stringfellow et al. 1997; Place et al. 2002). These groups have different functions ranging from production, marketing, and consumption. Working with groups is a more decentralized process and less top-down than working with individuals (Sanginga et al. 2001). Groups facilitate the organization of field days, promote efficient use of resources, improve farmers' collective confidence, and ensure that their needs are taken into account for the planning and implementation of programs. Farmer groups provide an opportunity to share ideas and labor and the exchange of information and thus create a multiplier effect, which facilitates the spread of relevant technologies (Mavedzenge et al. 1999).

Collective action through farmer groups may be aimed at different functions like generating, spreading, sharing, using, and applying knowledge and information. Different types of groups have developed in farmer-led research and extension.

Farmer groups in extension: Farmer groups are considered by both governments and donors as vehicles and entry points for new technologies and training for farmers. Extension becomes easy with involvement of these groups.

In the T&V system, 'contact groups' are groups of farmers providing a platform for interaction with public extension staff. For dissemination of technology from farmer to farmers groups play an important role. Farmer groups provide a tool to improve the cost efficiency of collaboration of researchers, farmers and extension workers as a result of logistical reasons (e.g. joint meetings, joint evaluations). Farmers are required to work in groups in the context of privatization and liberalization in agriculture.

Farmer groups are engaged primarily been in participatory technology development, participatory learning and action (PLA) research, and technology adaptation/dissemination. They take active part in technology dissemination and organization of farmer field schools and field days. Farmer to farmer extension is possible in group approach. The major functions of farmer groups are communication and information, dissemination of information and networking.

Participatory Extension Approach (PEA)

Participation is a process through which stakeholder's influence and share control over development initiatives and the decisions and resources that affect them (World Bank 1994). The participatory extension approach (PE) improves the effectiveness of rural extension through the participation of farmers. There are several reasons why it is desirable for farmers to participate in decisions regarding extension programs.

- They have the information which is crucial for extension planning.
- With shared responsibility they will be motivated to cooperate in the extension program.
- In a democratic society people have right to participate in decisions about the goals they hope to achieve.
- Many agriculture development problems can no longer be solved through individual decision-making and a collective decision would be required. This requires people's participation.

Over the last decade, there has been a growing awareness of the inadequacies of conventional approaches to agricultural research and extension, which necessitates a participatory extension.

The successful use of a (PEA) relies heavily on the motivation and ingenuity of the field workers. In top-down extension, the farmers' brains are rarely used, whereas in participatory extension all brains are used together and this is a two-way channel.

Principles of Participation

Pretty (1995) identifies seven clear types of participation such as passive participation, participation in information giving, participation of material incentives, participation by consultation, functional participation, interactive participation, and self- mobilization. The principles of participation include the following.

- **Inclusion:** All people or representatives of all groups who will be affected by the results of a decision or a process are to be included.
- **Equal partnership:** Every person has equal rights to participate in the process regardless of their status.
- **Transparency:** All participants must help create a climate conducive to open communication and building dialogue.
- **Sharing power:** The stakeholders need to share the authority and power.
- **Sharing responsibility:** All stakeholders have equal responsibility for a decision that is made and each should have clear responsibilities within each process.
- **Empowerment:** Participants with special skills should be encouraged to take responsibility for tasks within their specialty, but should also encourage others to also be involved to promote mutual learning and empowerment.
- **Cooperation:** Cooperation is very important as sharing everybody's strengths reduces everybody's weaknesses.

Rationale for Participatory Extension

The problems with the public extension system have been (i) poor response to farmers' needs, (ii) absence of ownership by the clients, (iii) failure to reach poor and women farmers;(iv) limitations in the quality of field and technical staff; and (v) high and unsustainable public costs. Some of these problems have been eased by modifying the T&V system by group extension and the use of multi-media. However, it is required to educate and enable the farmers to define and solve their own problems in order to meet the diverse needs of modern farming. There is a need to provide demand-based extension and cost-sharing between the public extension and farmers, and contract service to private firms, NGOs and/or cooperatives, farmers 'organizations, and self-help groups. In summary, the following requirements may go in favor of participatory extension.

- Participation favors sustainability because any action is based on the client's needs and interests.
- Participation tends to lead to a broad impact because extension contents and methods are suitably adapted and because of the direct interaction among the clients.
- Participation favors efficiency because minimum staff will be necessary for a broad impact.
- Participatory extension safeguards effective two-way communications among clients, between clients and other protagonists, and within its own structures.
- True participation depends on a lively partnership characterized by mutual trust.

Importance of Participatory Extension

Participatory extension provides a partnership between farmers, researchers, and extension workers to learn from each other. It is intended to promote farmers' ability to adapt and generate new technologies and practices. The extension worker works as a facilitator in the process of adult learning, group extension, farmer field school, farmer-to-farmer learning, extension program planning, diagnostic survey, demonstrations, PRA/ Rapid Rural Appraisal (RRA), etc. The participatory extension is a vehicle and Participatory Rural Appraisal is a toolbox. The importance of participatory extension is outlined below.

- It ensures to work cooperatively;
- Ensures learning through sharing;
- Help farmers to develop the ability to encourage each other in farm activities;
- Encourage to practice through various options;
- Mobilizes a large number of people;
- Make farmers free to ask for advice;
- Brings desirable techniques;
- Brings about development in the area; and
- Proves effectiveness in sustainable development.

Key Elements of Participatory Extension

Stakeholder commitment: Broad consultation from the outset is needed to ensure sufficient commitment to change by all stakeholder groups. Extension services that are participatory and accountable to farmers imply some loss of control for central planners at the government level.

The Institutional framework: There is no single institutional model for delivering participatory extension services. The formation of farmers' groups has facilitated the dissemination of agricultural innovations and provided all farmers with better access to farm inputs and credit. A multi-institutional approach recognizes that farmers get information from several different sources and that some organizations are more effective in reaching certain categories of farmers. Other key issues include: instituting incentives and mechanisms for accountability to farmers on the part of extension workers; identifying where legal and regulatory changes are needed; training staff in participatory methods; building the capacity of local farmers groups; and ensuring that local-level institutions do not exclude some groups of farmers from participation.

Two-way communication: The function of the extension is not confined to technology transfer. It ensures effective two-way flows of information, with the aim of empowering farmers through knowledge rather than giving only technical prescriptions (Figure 7.1). Tools for such two-way flow include beneficiary assessment, gender analysis, participatory rural appraisal, and problem analysis. Joint problem-solving and decision-making is achieved through workshops, round-tables, public hearings, and farmers' organizations. There is considerable potential in adapting the use of mass media and information technology to support participatory extension, channeling feedback from rural communities to researchers and extension workers, as well as providing information to farmers. Farmer participation in designing and implementing mass media programs improves program quality and enhances the learning process.

Figure 7.1: Participatory extension in the field.

Participatory Approaches

In this approach, the researchers, farmers, and extension workers work together and learn from each other. Participatory tools of RRA, Participatory Rural Appraisal (PRA), Participatory Needs Assessment (PNA), PLA, and Participatory Monitoring and Evaluation (PME) are used in this approach. The focus of the approach is on problem diagnosis, community empowerment, and farmer-led research and extension.

RRA, PRA, and PLA have been dealt in detail in Chapter 10 of this book while PME has been dealt in detail in Chapter 8.

Basic Features and Considerations of Participatory Approaches

The basic features of the participatory approach are the involvement of the whole community, triangulation of information for cross-checking, the leadership of the community/residents in all phases, self-help, mutual help, and changes in attitude and behavior.

The participatory approaches take into consideration the contribution of different socio-interest groups (women, men, and youth). The diagnosis needs to be facilitated by the multi-disciplinary team by discussion with different groups and the cross-checking of information may be made by triangulation.

The participatory approach emphasizes empowering farmers to implement their own decisions and a leadership role is required at the community level. The participation should be based on self-help and mutual help ideologies. All residents are to be given opportunities in planning, implement monitoring, and evaluation. The extension workers should not impose any idea on the community members as communities know their situation and problems better than outsiders do. The extension worker should facilitate the process and share his/her experience and knowledge with others.

General Principles of Participatory Extension Approach (PEA)

PEA follows certain principles during its planning and implementation. Group learning and community mobilization are necessary for increasing the strength of the poor by mobilizing them into groups and involving them in group analysis, and mobilizing them for planning and action, taking equal partnership, contributing knowledge and skills, and learning from each other.

The capacity building of the community is essential to identify problems, plan and manage actions to solve them; get involved in technology development; decide appropriate interventions, and take steps to attain sustainable growth.

PEA Cycle and the Key Stages

The PEA cycle follows a logical sequence of interactive stages. The action plans are evaluated after implementation and the diagnosis is updated to reflect the progress,

The PEA cycle follows a cycle of eight stages (Figure 7.2).

Stage 1: Preparation: Preparation involves various activities, such as training the core team of trainers, creating awareness among key stakeholders about the PEA cycle, building a multi-disciplinary district team, identifying of community in consultation with district leaders, and reaching the community for rapport building and explaining the purpose of the PEA process.

Stage 2: Diagnosis: Conducting a participatory diagnosis to examine the community, practices, opportunities, the problems they encounter, and their causes in PEA. The situation analysis is done by using PRA tools and techniques. Understanding the limitations and potential of an area, and the physical and social-economic landscape makes it easier to find appropriate solutions. This gives opportunities to residents including women and the poor to analyze their own situation with the facilitation of the multi-disciplinary team.

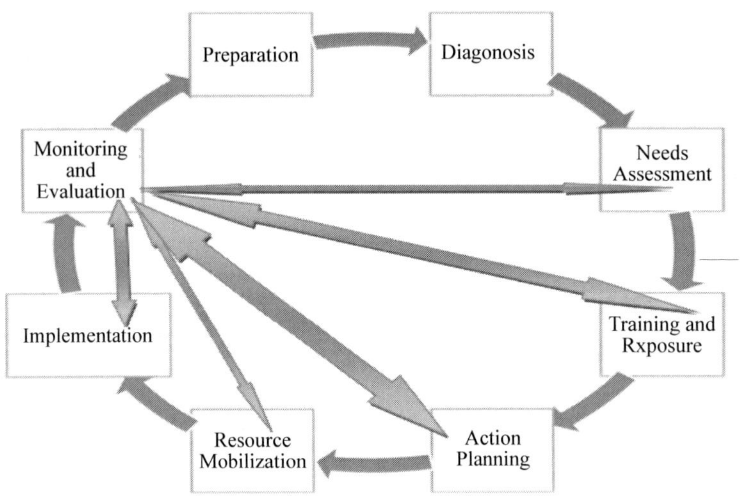

Figure 7.2: PEA facilitation cycle.

Stage 3: Needs Assessment: Before the assessment of community needs the interest group members are identified and their needs are assessed at the household level with a gender perspective and at the community level. The needs may include technical, information, management, exposure, financial, infrastructural, security, networks, and commercial needs.

Stage 4: Training and Exposure: The training need assessment is made using the required tools to build the capacity of the community to address felt needs identified under the needs assessment.

Stage 5: Action Planning: Action planning involves the preparation of community residents' proposed action plan, deciding on proper action plans and alternatives, revising the plan, and specifying the practical steps needed to implement the activities and the tasks and roles of all the partners involved in the process.

Stage 6: Resource Mobilization: Arrangement of cash, land, physical inputs, and labor for implementation and support sufficiently to source the logistics necessary to implement the action plan.

Stage 7: Implementation: During this stage, the community residents decide on the kind of organization they need to coordinate and implement the development activities. The roles of different stakeholders are also decided for the implementation of the plan.

Stage 8: Monitoring and Evaluation and Impact Assessment: Participatory M&E is to be done using indicators decided by the community. An annual impact assessment is made to measure the progress and the impact on the well-being of the community. The whole process is then reviewed, and new actions are planned.

A diagrammatic PEA cycle is displayed in Figure 7.2 and summarized in Table 7.1.

In practice, PEA follows four phases such as social mobilization, community-level action planning, implementation and trying out/farmer experimentation, and monitoring the process through sharing experiences ideas and self-evaluation as shown in (Figure 7.3)

Table 7.1: Phases of PEA

Phases	**Activities required**	**Tools to be used**
Preparation	Training of trainers, selecting communities, contacting authorities, awareness, meeting, and data collection	Secondary data collection/review
Diagnosis	Analysis of problem and solution by primary sources	PRA/PLA tools like participatory mapping, transect walks, vein diagrams, semi-structured interviews, trend analysis, seasonal calendars, ranking, scoring, problem cause, etc.

Assessment of community needs	Identification of various communities needs like technical information and infrastructure, etc.	SSI, Venn diagram, Ranking, etc.
Training and exposure needs	Training, exposure visits, study tours, etc.	Training needs assessment
Action Planning	Planning for individuals, groups, community action plans, and household action plans.	Solution-seeking, action-planning
Resource Mobilization	Listing resource requirements (land, inputs, finances, etc.)	SSI, FGD
Implementation	Taking action, monitoring, reporting/reviewing	
Monitoring & Evaluation	Reviewing progress, achievement, and impact	Progress reports, review meetings, participatory M&E, and third-party evaluation.

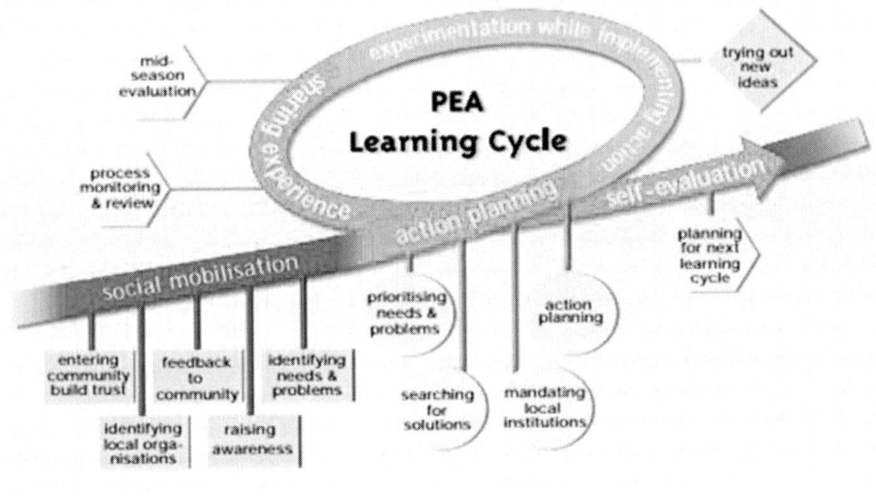

Figure 7.3: PEA learning cycle.

Benefits of PEA

The PEA has many benefits over the traditional approaches as outlined below.

- Generation of self-confidence and self-reliance to improve the standard of living;
- Community ownership for development and ensuring demand-driven services from the service providers;
- Participation in planning, implementation and M&E;

- Capacity development of all stakeholders;
- Strengthening of extension service delivery for increased production from crops and live stocks and farm income;
- Improving community resource utilization, poverty reduction, and sustainable rural development;
- Generation of data for future planning;
- Building a spirit of cooperation among the members;
- Developing new attitudes of all staff toward the organization and its managers and community members;
- Ensuring sustainability in rural development and making resources cost-effective; and
- Building capacity to adapt technologies and practices so as to make the most efficient use of their limited resources and their environment.

References

Gulati A, Sharma P, Samantara A and Terway P. 2018. Agriculture extension system in India: Review of current status, trends and the way forward. Indian Council for Research on International Economic Relations.

Kenmore, P. 2002. "Integrated Pest Management." International Journal of Occupational & Environmental Health 8, (3 (2002): 173–174.

Mavedzenge B.Z., Murimbarimba F., and Mudzivo, C. 1999. "Experiences of farmer participation in soil fertility research in southern Zimbabwe." Managing Africa's Soils 5.

Nagel U. J. 1980. Institutionalization of Knowledge Flows: An Analysis of the Extension Role of Two Agricultural Universities in India. Frankfurt (Main), Germany: DLG-Verlag, Frankfurt (Main), Germany.

Place F., Swallow B.M., Wangila J, and Barrett, C.B. 2002. "Lessons for natural resource management technology adoption and research." In: Natural resources management in African agriculture: Understanding and improving current practices, edited by Barrett, C.B.

Pretty, J. N. 1995. Regenerating Agriculture Policies and Practice for Sustainable and self-Reliance. (London: Eastern Publication Ltd and Washington, D.C: National Academy Press, 1995.

Put, M. 1998. Innocent Farmers? A Comparative Evaluation into a Government and an NGO Project Located in Semi-arid Andhra Pradesh, India, Meant to Induce Farmers to Adopt Innovations for Dryland Agriculture. Amsterdam, the Netherlands: Thela Publishers, Amsterdam, Netherlands.

Rogers, E. 1995: Diffusion of Innovations. Free Press.

Sanginga, P.C., Lilja, N., and Tumwine, J. 2001. Assessing the quality of participation in farmers' research groups in the Highlands of Kabale, Uganda. PRGA; CGIAR, Future Harvest. Working document/PRGA 19.

Stringfellow, R., Coulter, J., Lucey, T., McKone, C., and Hussain, A. 1997. Improving the Access of Smallholders to Agricultural Services in sub-Saharan Africa: Farmer Cooperation and the Role of the Donor Community. London, UK: Natural Resource Perspectives 20. Overseas Development Institute, London, UK.

World Bank. 1994. Agricultural Reform Implementation Support (ARIS) Project, Russian Federation. Staff Appraisal report, Agriculture, Industry and Finance Division, Europe and Central Asia Region. World Bank, Washington, DC, USA: World Bank.

Suggested Reading

Axinn, G.H. 1988. Guide on alternative extension approaches. Rome, Italy: FAO (Food and Agriculture Organization of the United Nations), Rome, Italy.

Babu, S C, Joshi P, Glendenning C J, Kwadwo A O and Rasheed S V. 2013. The state of agricultural extension reforms in India: Strategic priorities and policy options. Agricultural Economics Research Review: 26(347-2016–17086), 159.

Ban, A W Vanden 2006. Changing roles of Agricultural Extension in Asian Nations. Delhi: B.R. Publishing Corporation.

Desai, G R. 2004. Reforms in Extension: Experiences from ITD component of NATP in Orissa / Bhubaneswar: Institute on Management of Agricultural Extension.

Hansra, B S. 2001. Modernizing Indian Agriculture in 21st Century: Challenges, Opportunities and Strategies. New Delhi: Concept Publishing Company.

Kaegi, S. 2015. The experiences of India's agricultural extension system in reaching a large number of farmers with rural advisory services. Presented at the Background paper for Workshop Reaching the Millions at Hanoi, Vietnam.

Samantha, R K. 1997. Management of Agricultural Extension in Global Perspectives. Delhi: B.R. Publishing Corporation.

Scarborough, Vanessa 1997. Farmer-led extension: Concepts and practices. London: Intermediate Technology Publications.

8

Extension Program Planning Monitoring and Evaluation

Extension Program

Planning an extension program involves decision-making about the work of the organization. Extension program planning covers a set of objectives to be achieved by various activities included in the planning. Extension organization needs to decide the goal it aims at, the target group, the content of extension message, its use, and organization of all activities. The program includes situation analysis, and an action plan for solving people's problems which is drawn in advance. Extension workers have to decide the goals, target group, messages, extension methods, and media and organization of activities. While setting the goal, the perception of the present situation and the vision of the desired situation should be borne in mind. The target group is normally larger than the extension worker contacts directly or through mass media. The target group can be contacted indirectly. The extension content depends on the goal, the target group, and the extension strategy. The content should be so selected that it can be implemented with the resources available to the farmers locally. A combination of extension methods need to be used. The extension program needs to be flexible and can be easily implemented by the organization.

Extension Program Planning

Extension program planning indicates an account of the activities to be undertaken by the organization.

The essential phases of program planning in extension are analyzing the situation, setting goals and objectives, identifying problems and gaps, developing work plan, implementation, monitoring and evaluation, and reconsideration.

Steps in Extension Program Planning

Program planning in extension usually takes the following steps (Figure 8.1).

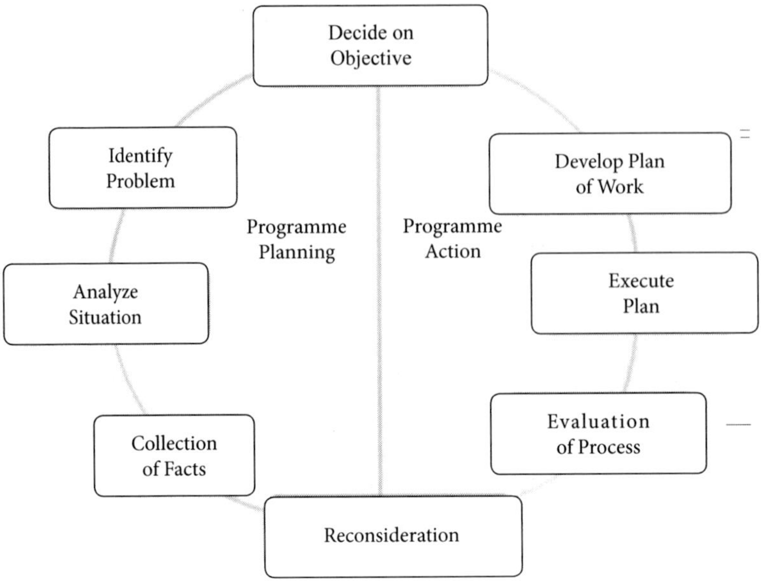

Figure 8.1: Extension program planning process.

Collection of facts: As a stock-taking exercise, relevant facts about the villages, people, physical conditions, existing agricultural practices, trends & outlook, customs, traditions, rural institutions, peoples' organizations operating in the area, etc. are collected from secondary sources and primary sources using participatory rural appraisal techniques.

Analysis of the situation: Situation analysis is made after analysis of data collected. This gives an indication of gaps found in the village and possibilities to overcome the gaps.

Problem analysis: The gap analysis also leads to problem analysis. The problems faced by the community need to be carefully analyzed along with steps already taken for their solution. The problems may have to be prioritized using PRA tools for the allocation of resources. This reflects people's need which needs to be addressed.

Setting goal and determination of objectives: Once the needs & problems of the people have been identified, it will be easier to set the goal and determine the various objectives, either short-term or long- term, which need to be planned for action. While setting goals, perception of the present situation, the vision, and the possibilities of bringing about a desirable change are to be considered. Under the hierarchy of goals, there may be several intermediary goals. For example, the ultimate goal may be improved welfare of the citizens, while the intermediary goals may be higher income from farms, higher productivity of farms, higher productivity of milk, improved crop production practices, improved livestock production system, etc.

Developing Plan of Work: A work plan is developed based on the goals & objectives of the extension program, and resource availability. The plan of action clearly indicates the activities, roles and responsibilities of the implementers and the time limit for the accomplishment of the action points.

Implementation of the Plan of Work: After the plan is approved necessary arrangements are made for the supply of required inputs, credits, teaching aids, extension literature, etc. for implementation of the plan. The implementation is done in a participatory manner by using extension methods with active involvement of people.

Monitoring & Evaluation: Monitoring of the program is done at suitable intervals to take up mid-term corrections. The evaluation is done to measure the physical progress, efficiency, and effectiveness of the methods and techniques used and the planning process. An extension program evaluation means determining the results of extension programs with regard to the goals and objectives and deciding the changes required in consequent planning.

Reconsideration: Based on the results of the evaluation, the program may be reconsidered, if necessary, to make it more effective and sound.

Rationale of Program Planning

Extension program planning is a blue-print for action that gives direction and ensures effective learning. Kelsey and Hearne (1949) have identified the following reasons for planning extension programs.

- Program planning is based on situation analysis;
- Selects problems based on needs;
- Defines objectives of the programs;
- Reflects performance with flexibility;
- Incorporates balance with emphasis;
- Envisages a definite plan of work is a continuous process;
- Is a teaching process;
- Is a coordinating process;
- Involves local people and their institutions; and
- Provides scope for evaluation of results.

Extension Program Planning Model

Sandhu (1965) developed a model for planning extension and rural development programs with six phases (Figure 8.2)

Organization for planning: In an extension program planning people are involved from the beginning in the participatory situation analysis. It is necessary that the target groups need to be identified and members of the planning committee should be represented by all such social groups.

Planning process: The extension staff and people's representatives are required to be acquainted with the principles and procedures of program planning for clarity and uniform action. The extension workers and planning committee members should have a clear understanding of their roles, responsibilities, purpose, and scope of planning including the process and time schedule. The steps are analyzing the situation by collecting facts and analysis of data, determining the objectives, analysis of problems and needs and their prioritization, and working out solutions for identification and implementation.

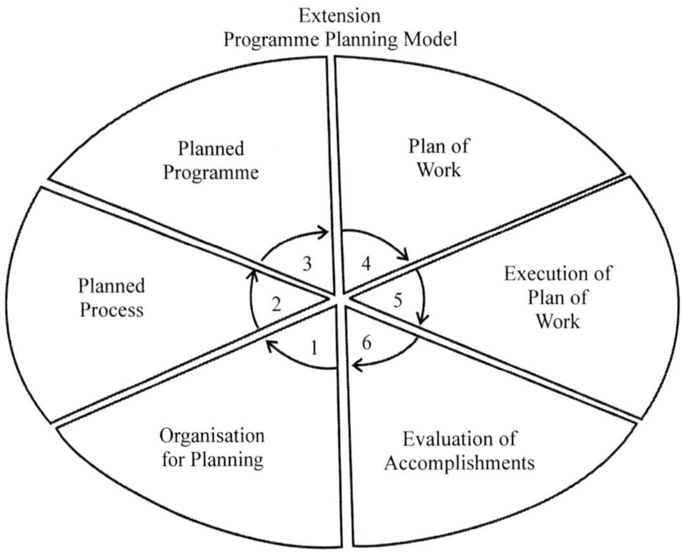

Figure 8.2: Extension program model.

Planned program: The plan is a blueprint of action where goals and objectives are clearly written and can be understood by the stakeholders and get approval. The annual work plan can be drawn from the planned program.

Plan of work: Plan of work is a list of activities to achieve the objectives which are developed by the extension workers in association with the stakeholders. It indicates the roles and responsibilities of the stakeholders. The plan indicates all details of work to be done with the time schedule and roles of the stakeholders in its implementation.

Implementation: Advanced arrangement of key inputs, and teaching materials, and carrying out all planned activities are undertaken in the implementation phase. The requirement of inputs will be calculated by the extension workers and the stakeholders and staked in advance. The teaching aids should be prepared in advance by the extension workers. The approved plan should be explained clearly to the stakeholders and people's representatives before implementation in a coordinated manner. The activities are to be carried out as per the calendar with shared responsibilities.

Evaluation of accomplishment: Any program requires a concurrent and *ex-facto* review to assess the success or failure. Evaluation is necessary to justify the appropriation of public funds and continuing support from the people.

Monitoring of Extension Plans & Programs

Monitoring: Monitoring primarily provides information on project performance and gives signals on whether the activities are carried out according to the plan. It also provides information on the socioeconomic indicators for ex-post evaluation. It can be done through review meetings and periodic seminars which give direction to the management to compare the progress of work against planned activities, detect deviations, identify bottlenecks, and take corrective actions Monitoring is a continuous/periodic review by the project management staff, at every level of the implementation of an activity to ensure that input deliveries, services are carried out according to the plan to achieve targeted outputs. According to Bamberger (1986), it is an internal project activity designed to provide constant feedback on the progress of a project, the problem it is facing, and the efficiency with which it is implemented.

Purpose and Objectives of Monitoring

The purpose and objectives of monitoring are listed in Table 8.1.

Basic Concepts of Monitoring

The capability, effectiveness, and efficiency of a program/project are monitored while the impact is measured by evaluation. Monitoring of programs/projects is required to check the progress, resource utilization, assess the result and decide the mid-term corrections. In principle, monitoring should be simple, timely, relevant, participatory, flexible, action-oriented, and cost-effective.

Monitoring may be beneficiary contact monitoring, process monitoring, and physical and financial monitoring. Process monitoring as an approach is adopted to ensure that the processes are steered to achieve the desired results.

Monitoring could be undertaken by reviewing the regular progress reports submitted by the field staff which can indicate the progress of resource utilization and physical progress against the target. Monitoring staff performance can ensure that individuals are effectively employed to fulfill their tasks. This can be assessed in regular staff meetings. The higher officers can check the quantitative and qualitative aspects of the progress during field tours. The reports of visitors can give an indication of the progress of the project with deficiencies for correction. Participatory monitoring is a new technique where the involvement of the project staff and beneficiaries takes place.

Generally, quantitative and qualitative indicators are used in monitoring (Table 8.2). Quantitative indicators are measurable and can be direct indicators, indirect indicators, process indicators, and progress indicators. Monitoring in extension uses some special indicators like extension capability indicators and extension performance indicators in addition to the above indicators. In order to assess the status of the extension's capability at a certain time and determine the changes in it over time, capability monitoring is necessary. It can be done by desk review using in-house data annually.

Table 8.1: Purpose and Objectives of Monitoring

What	Review of the progress of program implementation with reference to action.
Why	To identify shortfalls, deviations, problems, and reasons of thereof for taking appropriate corrective/ remedial action to keep the project on track.
When	During the implementation of the project from inception till completion of the project.
Who	The project/ program management team and the beneficiaries

Table 8.2: Qualitative and Quantitative Indicators in Monitoring

Type of indicator	Indicators	Description
Quantitative	Direct	Measurable indicators like crop yield and profit.
	Indirect	These are proxy indicators that serve as substitutes.
	Process	Procedure in planning, designing, collecting funds, and construction of infrastructure are process indicators.
	Progress	Measuring progress against the target are progress indicators which are mostly quantitative.
Qualitative		Statement about the processes adopted and outcomes.

Extension performance indicator refers to the level of achievement of project targets and can be grouped as effectiveness indicators and project performance indicators. Effectiveness indicators could be single or composite. Project impact is the effects of outcomes on rural people as a whole. It can be indicated by changes in levels and distribution of project benefits.

Conventional progress monitoring focuses on physical, financial, and logistical aspects of projects whereas process monitoring deals with critical processes which are directly related to the project objectives.

Evaluation

Evaluation in extension is required to determine whether the extension program has achieved the goals and whether these goals could have been achieved more effectively in a better way (Figure 8.3). It provides a learning experience for the extension officers. The experience gained could be utilized while preparing the next extension programs. Hence, it is used to improve present and future activities such as planning, programming, decision-making, and implementation to achieve extension policy goals more effectively. Evaluation provides information about past or ongoing activities as a basis for modifying or redesigning future strategies. Evaluation plays the same role as feedback in the communication process. Evaluation normally makes an assessment of results achieved using certain criteria (Table 8.3).

Types of Evaluation

Evaluation could be ex-ante evaluation, ongoing evaluation, ex-post evaluation, and impact evaluation.

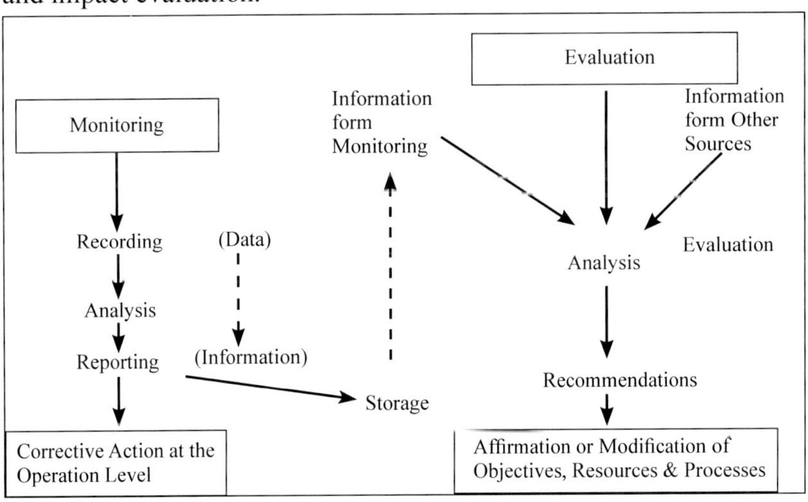

Figure 8.3: Relationship between monitoring and evaluation.

Table 8.3: Difference between Monitoring and Evaluation

Particulars	Monitoring	Evaluation
Period	Continuous: Stars and ends with a program	One-shot operation: At a point in time (mid-term or terminal)
Use	Required for immediate use and mid-term corrections	Used for future plan replication/ expansion
Agency	Done by implementing personnel	Usually by an outside agency
Coverage	Quick, but covers all units	In-depth coverage of a sample
Purpose	Correcting/managing	Learning process
Study	Symptomatic, early warning system	Diagnostic

Ex-ante evaluation: It includes a comprehensive analysis of the potential impact of alternative activities before the implementation of a project. The methods used in such evaluation are expert reviews using checklists, scoring models, and even cost–benefit analysis.

On-going evaluation: In this evaluation, the ongoing activities are reviewed at critical stages to determine if they should be continued, modified, or aborted. The ongoing evaluation addresses problems associated with the day-to-day management of interventions and also can indicate the need for changes in project objectives and targets. This type of evaluation is used to obtain feedback from the target group which is done through a series of meetings at the site with peers, farmers, extension staff, and NGOs.

Ex-post evaluation: This is a terminal evaluation, which assesses the project's performance, quality, and relevance immediately after the project completion. It measures the effectiveness and efficiency of the completed activities. It is linked to ex-ante evaluation and is best done against a baseline study. Ex-post evaluation analyses the project from beginning to end, assesses the project activities, and identifies causes for discrepancies, costs, and the quality and relevance of the activities against the project objectives. The methods usually used in this evaluation are statistical evaluation, economic evaluation, agronomic assessment, and farmers/community assessment.

Impact evaluation: Impact evaluation attempts to measure the contribution of the project to development goals, such as increased farm production, improved food security, poverty alleviation, etc. Impact evaluation must distinguish between the contributions of various sectors like research, extension services, agricultural inputs, adequate infrastructure, and favorable marketing and pricing policies make to national development.

Evaluation could be self-evaluation, internal evaluation, or external evaluation. Self-evaluation is undertaken by every extension worker as a matter of routine with a self-critical attitude. Internal evaluation is undertaken by the agency responsible for the planning and implementation. External evaluation is done by a third party to avoid bias in finding the results. Evaluation could be concurrent or end-term evaluation.

Evaluation could be formative or summative. Formative evaluation gathers data for the development of a program while summative evaluation gathers data on the results of the program in order to decide whether or not the program should be continued, expanded, or discontinued. Interest in formative evaluation is growing in recent years. It is mostly intended development of extension programs. Pre-testing is one form of formative evaluation that can provide valuable information on how to improve extension materials. When the goals differ with members of a community and there is process orientation, the summative evaluation becomes difficult. The target group may be asked about the extent to which they have participated in the program and how this has helped them achieve their goals.

Impact: The typical impact chain starts from the set of inputs and activities of a project to the output, outcome, and impact. The output, outcome, and impact are generally sequentially produced over a period of time. Output is the goods and services produced by a set of activities. The outcome is the first level effect of the outputs which may be measured in terms of behavioral changes. This refers to the results from the application of the output. In the case of technology adoption the intermediate outcome may be the effects at the farm/household level, i.e. increased yield and reduction in costs. The impact is the measurable effects of the outputs and outcomes on the well-being of the ultimate beneficiaries Very often the ultimate outcomes are closely linked to the sectoral/regional/national developmental goals. The impact may be positive or negative. The impact may be production impact, economic impact, socio-cultural impact, environmental impact, and institutional impact (Figure 8.4).

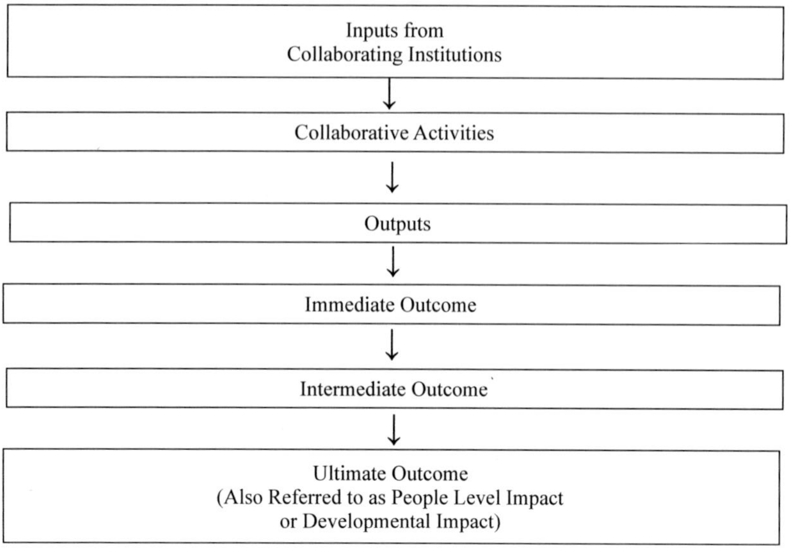

Figure 8.4: Impact chain or result chain.

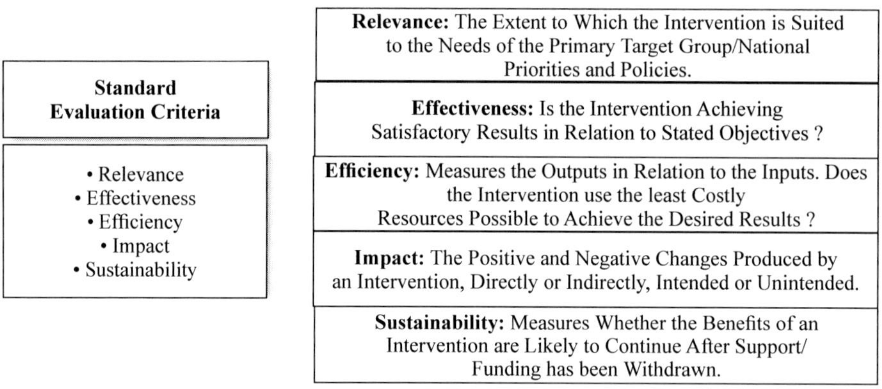

Figure 8.5: Standard evaluation criteria.

Criteria for Evaluation

The criteria for evaluation include clearly defined objectives, a valid instrument for measurement, objectivity – free from biases, reliability, accurate evidence for change, and practicability. The normal criteria used for the evaluation of projects are relevance, effectiveness, efficiency, impact, and sustainability as indicated in Figure 8.5.

Narayan (1993) differentiates between conventional and participatory evaluation in Table 8.4.

Extension evaluation process

When inputs are used in a program, some outputs are expected. The outputs may lead to the achievement of purposes and then realization of goals. Hence, evaluation should measure the outputs, outcomes, and their impact on the program. Evaluation of extension programs follows certain steps. These steps are (i) formulation of evaluation objectives, (ii) classifying program objectives, (iii) identifying indicators, (iv) deciding the kind of information needed, (v) sampling, (vi) design of the evaluation, (vii) collection and evaluation of evidence, and (viii) interpretation of results. The evaluation results indicate the achievements and failures, if any, so as to make future adjustments. The evaluation report needs to be made available to all concerned.

Table 8.4: Conventional and Participatory Evaluation

	Conventional evaluation	**Participatory evaluation**
Who	External experts	Community members, facilitators, and project staff
What	Predetermined indicators of success, principally cost, and production outputs	People identify their own indicators of success, which may include production outputs
How	Focus on scientific objectivity distancing of evaluators from other participants, uniform, complex procedures, delayed, and limited access to results	Self-evaluation, simple methods adopted to local culture, open, immediate sharing of results through local involvement in evaluation processes
When	Usually on completion of project/ program, sometimes also mid-term	More frequent, small-scale evaluations
Why	Accountability, usually summative, to determine if funding continues	To empower local people to initiate, control, and take corrective actions

There are some methods usually used in the evaluation of extension programs. They are:

- Single or multi-perspective view maintained.
- Statistical inferences made.
- Use of direct quotations from stakeholders.
- Use of a structured questionnaire.
- Use of focus-group discussions.
- Use of rapid rural appraisal and appropriate medium of inquiry.
- Application of experimental design.

The Logical Framework (LFA)

Log frames are used to identify the essential elements of a project and the linkages between these in a logical, concise, and objective manner. The framework serves as a tool for logically identifying inputs, outputs, success indicators, and assumptions for monitoring progress and evaluating performance. A log frame comprises a hierarchy of objective statements regarding the Goal, Purpose, Outputs, and Inputs (which include activities).

- **Goal**: The overall development aim (or vision) of any project is the goal. Usually, there is one broad goal, which is related to a wider national or regional objective. The project would not be expected to achieve this alone but to contribute towards this goal.
- **Purpose**: It expresses the primary reason for the project which can address certain problems and achieve development expected during the project or the situation envisaged at its completion.
- **Outputs**: Project outputs are the direct, identifiable, and measurable results expected from the use of inputs.
- **Inputs**: These are the resources made available, which together with the activities lead to the achievement of outputs. Inputs may be people, equipment, or finance.
- **Achievement**: Project activities are the tasks undertaken using resource inputs to achieve defined outputs. Observation of the execution of activities forms a basis for monitoring performance.

A log-frame matrix is prepared at the time of design of the project which is used at the time of evaluation (Table 8.5).

Table 8.5: Log-frame Matrix

	Narrative summary	**Measurable indicators**	**Means of verification**	**Assumptions**
Goal	What wider issue will the project help to resolve (e.g. a national objective)?	What are the indicators of goal achievement?	What are the sources of information? What methods are to be used to obtain it?	**(Purpose to goal)** What assumptions/ external factors must be true if the purpose is to help reach the Goal?

Purpose	What immediate changes/ benefits will the project bring about for the target population or area?	What are the measures to judge the project's immediate effects, benefits, and losses?	What are the sources of information? What methods are to be used for obtaining it?	**(Outputs to purpose)** What external factors must be true if outputs are to achieve the purpose
Output	What direct outputs will the project produce so it can achieve its purpose?	What kind and quantity of outputs are planned, and by when will they be produced?	What are the sources of information? What methods are to be used for obtaining it?	**(Inputs to output)** What external factors must be realized to produce the planned outputs in time?
Activities	What activities must be undertaken to produce the outputs? When must these activities take place?	**Inputs/ resources** What materials, equipment, or services are required, at what cost, and over what period?	What are the sources of information on inputs? What methods are to be used to obtain it?	**(Initial assumptions)** What actions outside the control of the donor are necessary to begin the project?

Source: Modified from AIDAB (1991) and Farrington and Nelson (1997)

In designing a project, planners must address both the vertical and horizontal logic of the log-frame. The vertical logic uses the principle of causality between levels, while the horizontal logic identifies and states the means by which the project will produce results at the various levels. Verifiable indicators are stated in the second column, and the method of collecting the information is recorded in the third column.

Participatory M&E

Participatory M&E is undertaken with the active participation of the stakeholders. PME aims at engaging and empowering communities through shared control, joint learning, and ongoing reflection to improve progress and achieve desired goals. It is an ongoing and regular process that actively involves stakeholders in all stages of collecting, analyzing, and using the information on intervention with a view to assessing the processes and results and making recommendations. Groups of local people actively participate in discussions and decide how they can best improve them. Different tools like participatory maps, group meetings, drawings/discussions, murals and posters, open-ended stories, most significant change stories, semi-structural interviews, survival surveys, posters, etc. are used in PME. (Basu et al. 2006).

Purpose

Participatory M&E ensures the participation of the stakeholders and enhances efficiency and effectiveness. It can contribute to results-based management by improving policymaking, facilitating adaptive management, enhancing the efficiency of resource use, and promoting staff motivation. It is based on openness, transparency, responsiveness, predictability, and accountability. PME enhances equity of outcomes due to the commitment of all stakeholder groups to question the existing distribution of services. Well-structured PME systems may help communities and civil society organizations to develop partnerships with projects, office bearers, and other stakeholders (Hilhorst and Guijt 2006).

Principles

The four broad principles of PME are participation, inclusiveness, learning, and flexibility. PME involves all stakeholders uses both quantitative and qualitative indicators acceptable to all, and gathers two-way information. PME emphasizes on beneficiaries and stakeholders not only as information providers but also as participants in the process of evaluation. Supplementing more formal methods of inquiry, such as standard questionnaires or one-to-one interviews, with non-formal techniques will enhance the quality of information. Experience suggests that it is preferable to start PME on a small scale and create opportunities for PM&E to be tested before the process is scaled up and introduced more widely. The stakeholders should contribute to designing the M&E process and agree to analyze the data together with the evaluator. The lessons learned will help improve the program further. An essential element of PM&E is that the numbers, roles, and skills of stakeholders, the external environment, and other factors are fluid and can change frequently (UNESCO 2007; Shah, 2017).

Process

The participatory monitoring and evaluation process includes four steps such as planning process, gathering data, data analysis, and sharing of information & defining actions to be taken (Aubel 2004) as displayed in Figure 8.6 and Table 8.6.

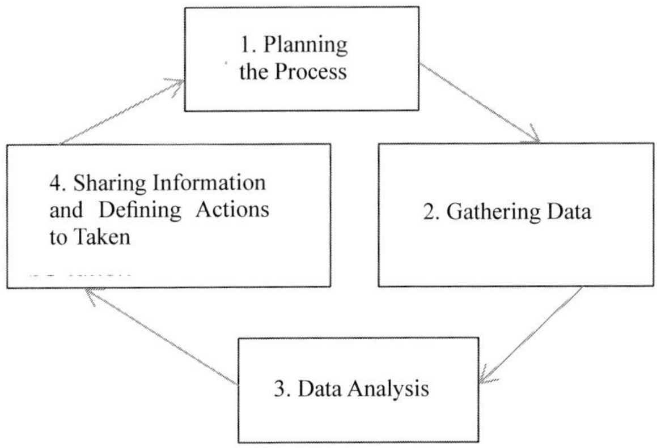

Figure 8.6: The planning process of PME (Aubel 2004).

Table 8.6: The planning process of PME

Step	**Process**
Planning process	The objective of the step is to include stakeholders for whom the development intervention is planned for.
Gathering data	Once the participants are decided, then methods like structured interviews (survey), group discussions, mapping, etc. are used to collect data.
Data analysis	In this step analysis of data collected is made.
Sharing information and defining actions to be taken	The information is shared with all community members, to define the actions to be taken.

Benefits of Participatory M&E (PME)

- PME allows a better understanding of the perspectives of community members, which can contribute to improved program implementation.
- It ensures the involvement of stakeholders, allowing them to reflect on their own experiences and to learn from them.
- Enhances reliability and gathers feedback for future correction.
- Increases the capacity and confidence of local program staff and community members to analyze their own needs and programs, and to undertake actions.
- Involvement of community and program stakeholders encourages articulation of community members their priorities and criticisms of development program strategies.

- Contributes to the sustainability of program strategies by increasing the sense of ownership of the community members.
- Increases the motivation of stakeholders to contribute ideas to corrective actions in the program.

Participatory Impact Monitoring (PIM)

Participation is required during the entire project cycle starting from need assessment, problem identification, project design, implementation, monitoring, and evaluation. The logical framework approach (LFA) is often used to identify objectively verifiable indicators for this purpose. Participatory impact monitoring (PIM) is a method to assess the socioeconomic impact of a project/ program on the target beneficiaries. PIM is defined as a method that is used to evaluate the socio-cultural impact of the project. Stakeholders like farmer groups, self-help groups, development organizations, NGOs, and the funding agency are involved in PIM. PIM is based on the joint perception of impacts by the stakeholders. PIM is usually conducted by all levels of stakeholders in the form of groups. Regular group meetings, the interest of members, the willingness of group leaders to communicate, and group members' willingness to invest a little time in joint management are required for PIM. In PIM different groups regularly exchange their information, perceptions, and interpretations at Joint Reflection Workshops. "What should be monitored, how it can be monitored, who should monitor, how results can be documented, how the results are analyzed, and what action should be taken" are the basic questions for PIM. The Joint Reflection Workshop is useful to meet every three months in the first year and annual meetings are required after the first year.

References

Aubel, J. 2004. Participatory monitoring and evaluation for hygiene improvement, beyond the toolbox: What else is required for effective PM&E? Retrieved from http://www.ehproject.org/PDF/Strategic_papers/SR-9%20Lit%20Rev.pdf on 5-11-2018.

Bamberger, M. 1986. World Bank technical paper, Monitoring and evaluation of urban development programmes. Reteived from www.fao.org.

Basu, D., Adhikary, M.M., and Biswas, D. 2006. Participatory Monitoring and Evaluation of Development Programs: Practitioner's Guide. Udaipur: Agrotech Publishing Academy, Udaipur.

Hilhorst, T. and Guijt, I. 2006. "Participatory monitoring and evaluation - A process to support Governance: A guidance paper", report to the World Bank. Amsterdam: Royal Tropical Institute (KIT), Amsterdam.

Narayan, D. 1993, Participatory Evaluation: Tools for Managing Change in Water and Sanitation, World Bank Technical Paper Number 207.

Sandhu, A.S. 1965. "An Appraisal of the Procedures Followed in Planning Agricultural Extension Program at the Block Level in Punjab." M.Sc. Thesis. PAU, Ludhiana.

UNESCO 2007. A guide for monitoring and evaluation community based projects. Retrieved from http://www.bibalex.org/Search4Dev /files/282315/114599.pdf on 20-11-2018.

Suggested Reading

Ahmed, M. R. 1991. Planning and Designing Social Forestry Project. In Ahmed, M. R. (ed.) Extension Teaching Methods, The Communication Process and Program Planning to Meet People's Needs in Extension Education in Community Development.

Farrington, J., and Nelson, J. 1997. "Using Log frames to Monitor and Review Farmer Participatory Research". Network Paper No. 73, London: Overseas Development Institute, London.

Jessica, Dart, R. John Petheram, Warren Straw S. 1998. Institute of Land and Food Resources, Vic "Review of Evaluation in Agricultural Extension" Publication No 98/136 Project No VCA-3A.

Loveleen, Kaur, and Sukhdeep, Kaur K. 2019. "Participatory monitoring and evaluation in extension programs": Asian Journal of Home Science Volume 14, 2 (2019): | Issue 2 | December, 2019 | 458-462.

Sandhu A. S. 2003. Extension Programme Planning. New Delhi, Oxford and IBH Publishers.

Swanson BE, Bentz RP and Sofranko AJ. (eds). 1997. Improving agricultural extension: A reference manual. 3rd ed. FAO (Food and Agriculture Organization of the United Nations), Rome, Italy.

USAID. 1973. The Logical Framework: Modifications Based on Experience., Washington, DC: United States Agency for International Development, Washington D.C.

Van den Ban, A. W. and Hawkins, H. S. 1996. (2nd Edition) Agricultural Extension, Blackwell Science, Berlin, Germany.

9

Role of Extension in Community Development and Rural Development in India

Concept of Community Development

A community is a group of people, who live in a geographical area and have interests in each other for the purpose of making a living. It is a form of social organization existing between the family and the state. Development connotes growth and implies gradual and sequential phases of change. A community development program means a program for gradual change in a group of people living in a geographical area and having an interest in each other for making a living. It involves the principles of sustainable development, empowerment, inclusivity, social justice, human rights, participative democracy, and equality. It is a collaborative and facilitative process of people who share a common purpose of building the capacity to have a positive impact on quality of life.

- Community development is a movement intended to promote better living for the whole community with the active participation and on the initiative of the community.
- It is a balanced program for stimulating the local potential for growth in every direction in both wealth and welfare by themselves with minimum outside help.
- It is technically aided and locally organized by self-help.
- Community development is adopted by many governments to reach village people and to make more effective use of local initiative and energy for increased production and better living standards.
- Community development is a process of social action in which the people of a community organize themselves define their needs and problems and plan for action.

Community Development Programs (CDP)

The community development program (CDP) is a multi-project program and it covers agriculture, animal husbandry, irrigation, cooperation, public health, education, social education, communication, village industries, etc. at the Block level. Extension officers for each sector were posted at the district level to plan, execute, and evaluate the program up to the village level. It was a country-wide program for rural development through trained extension officers by promoting local leadership and youth programs.

Organization

The program runs from the national level through state, district, and block levels to the village level to reach the community. Specific administrative officers were posted at the state, district, and Block levels, who worked under the Panchayati Raj System and district, Block, and Village levels. The Block Development Officer worked at the Block level under the administrative control of the District Collector/Deputy Commissioner and supported by the extension officers from development areas. At the Panchayat level multi-purpose Village Level Workers were posted for the extension work under the guidance of the Block level Extension Officers.

Philosophy: The basic philosophy of the program was based on the "felt needs" of people who want to be free from poverty with freedom in controlling their own socio-economic, religious and education development by self-help. The program aims to bring about changes in attitude, skill, habits, level of knowledge and intellectual advancement of people.

Approach: The various approaches used in community development include: using an asset-based approach that builds on strengths and existing resources; encouraging inclusive processes that embrace community diversity; and community ownership through collaboratively planned and led initiatives.

Aims: Community development aims at enabling people to collectively identify, prioritize and plan to solve their problems that ultimately lead to their sustainable development.

Community development values: Community development values are social justice, participation, equality, learning, and co-operation. It enables people to claim their human rights; democratic involvement people in the issues, that affect their lives; claim equal rights without any discrimination; and gain the skills, knowledge, and expertise that people contribute and develop by taking action to tackle social, economic, political, and environmental problems in a co-operative manner.

Community Development Program (CDP) in India: Based on the success of the Etawah Project in 1948, Community Development in India started on October 2nd, 1952 to bring an integrated development of rural people.

Community Development Program (CDP) has been divided into three phases: (i) the National Extension Phase; (ii) the Intensive Community Development Project Phase; and (iii) the Post-Intensive Development Phase.

In the first phase, the selected areas were subjected to the method used for providing the services based on the pattern of ordinary rural development, and that too with less government expenditure. In the second phase, the selected blocks were subjected to more composite, and the development schemes were more intensive and had a much larger expenditure. The third phase presumed that the basis of the process initiated for self-perpetuation was created during the previous phases along with a reduction in the required expenses of the special government. As per Balwantrai Team suggestions some modifications were made in which there was only one single scheme for the development for a span of 10 years in two stages. Even after this, the blocks were not abandoned and continued till the complete reconstruction was achieved. The objectives of the CDP were to improve the living standards of rural people promote economic development, and adopt a bottom-up approach with the active involvement of villagers in the planning and implementation of the program.

The CDP was implemented in three stages for the identification of villages, formation of Panchayats, and implementation of the projects. The program achieved some successes, in the construction of roads, schools, and irrigation canals.

Both the CDP, 1952 and the National Extension Service, 1953 worked for the development of agriculture, animal husbandry, roads, health, education, housing, employment, and social and cultural activities.

The Block Development Officer (BDO) was responsible for each Block and was assisted by a team of extension officers and Village Level Workers (VLW).

Based on the recommendations of Balwantrai Meheta Study Team (1957), Panchayati Raj was launched on October 2nd, 1959. The CDP was streamlined with democratic decentralization. Panchayati Raj provided for a three-tier structure with the Gram Panchayat at the village level, the Panchayat Samiti or intermediate Panchayat at the Block/Taluk level, and Zilla Parishad or Zilla Panchayat at the district level.

Program activities of CDP

The main items of program were improvement in farming techniques to boost production, development of cottage industries for employment, extending irrigation facilities, improvement in transport, education, health and sanitation, and cooperation at the Panchayat level.

Extension and community organization were important areas of CDP. Extension Education imparted proper knowledge, insight, and desire to improve the conditions of living of rural people. For Community Organization, there were three basic institutions namely, the village Panchayat, the village Cooperative, and the village school with functions of a development authority, economic development, and intellectual sphere, respectively. There were many community organizations that were linked to Panchayats for development works.

Problems with CDP

Continuous evaluation of CDPs has indicated some weaknesses in CDP:

- Insufficient attention to agricultural development with poor agricultural techniques;
- Slow development of irrigation and lack of irrigation in many blocks;
- Slow progress in animal husbandry;
- Unsatisfactory progress in education and health;
- Disparity in benefits and uneven progress;
- Unsatisfactory functioning of Panchayats;
- Lack of coordination;
- No contribution of community members;
- Financial stringency; and
- Slow rate of progress.

Suggestions for improvement

Balwant Rai Meheta committee and CD Mission made the following recommendations for improvement of CDP in India.

- Democratic decentralization at three tiers of Panchayati Raj system.
- Priority in the drinking water, irrigation, agriculture, and allied sectors.
- Collective effort of development and benefits.
- Strengthening of staff.
- Promotion of village industries.
- Establishment of service co-operatives.
- Greater role of Panchayat in communication, coordination, and use of unskilled labor.

Panchayati Raj

Balawantrai Committee (1957) recommended a three-tier Panchayati Raj system in India having a Zilla Parishad at the District level, Panchayat Samiti at the

Block/Takuk level and Gram Panchayat at the village level. Subsequently Ashok Meheta Committee (1978) recommended a two-tire Panchayati Raj at the district level and at the Mandal level. LM Singvi Committee recommended that Gram Sabha should be the base of decentralization which led to the Constitutional Amendment Act, 1993. With the 73rd and 74th Constitutional Amendments, significant powers were given to Panchayats and Municipalities. As per Article 243G of the Constitution, the Panchayats are expected to prepare plans for economic development and social justice including 29 subjects listed in the 11th Schedule of the Constitution. According to Article 243-A of the Constitution, a Gram Sabha may exercise such powers and perform such functions at the village level as the Legislature of the State may, by law, provide.

Panchayat Extension to the Scheduled Areas (PESA) Act, 1996

After the introduction of the 73rd Constitution Amendment Act, 1993, a restructuring of governance and administrative system was initiated in the country based on the philosophy of decentralization. Realizing that the tribal communities are most marginalized and remain isolated from the mainstream of development, the Working Group (1996) of the Ninth Five Year Plan suggested participatory planning for growth and equity in tribal areas. In order to strengthen the grassroots-level local bodies Panchayat Extension to Scheduled V Areas Act (PESA) 1996 was passed for the self-rule of tribals. Under PESA every Gram Sabha shall be competent to safeguard and preserve the traditions and customs of the tribes, their cultural identity, community resources, and customary mode of dispute resolution. Extensive powers have been given to Gram Sabha for approval of plans and programs and projects for socio-economic development and their implementation, identification of beneficiaries for poverty eradication programs and certification of utilization of funds by the Panchayat. Gram Sabha or Panchayat at the appropriate level has been given extra powers as indicated below.

- To be consulted in the matters of land acquisition and resettlement.
- Grant of license for mining leasing of m in or minerals and concessions for such activities.
- Planning and management of minor water bodies.
- Enforce prohibition or regulate or restrict the sale and consumption of any intoxicant.
- Ownership of minor forest produces.
- Prevent alienation of land and restore any unlawfully alienated land of STs.
- Mange village markets.
- Exercise control over money lending to STs.

- Exercise control over institutions and functions in all social sectors.
- Control local plans and resources.

The States were suggested to amend their respective Panchayat Acts to extend the provisions to Scheduled areas and make Rules for the implementation of PESA in their states.

Decentralization in Planning

Part IX was inserted by the Constitution (73rd Amendment) Act, 1993 to make Panchayats responsible for all development activities. The Panchayats were devolved with financial, function, and functionary powers for planning for economic development and social justice and implementation. All Panchayats are given financial assistance from the Central Finance Commission (CFC) and State Finance Commission (SFC) in addition to their own power of revenue generation. With the Constitution Amendment, the following mandatory provisions were made.

- Establishment of State Election Commission (Article 243 K)
- Holding regular Panchayat elections (Article 243 E)
- Reservation of seats for SC/ST and women (Article 243 D)
- Establishment of State Finance Commission at regular intervals (Article 243-I)
- Setting up District Planning Committee (Article 243 ZD)

The CFC grants are provided as basic and performance grants. Basic grants are meant for the delivery of basic services like water supply, sanitation, sewerage and solid waste management, stormwater management, maintenance of community assets, roads, footpaths, street lights, and burial and cremation grounds. It is provided with 90% weightage to the 2011 population and 10% weightage to the area. Performance grants are allowed for ensuring reliable auditing accounts and data of receipts and expenditures and improvement of own revenues. State-specific guidelines are released to prepare the Gram Panchayat Development Plan (GPDP).

In decentralized planning, real impetus was given only after incorporating a provision in article 243ZD in the Constitution of India. Both the 73rd and 74th Amendments (1992) to the Constitution were passed and provided opportunities to the local rural and urban bodies for preparing micro plan from the villages and urban local bodies to the district in a decentralized manner. Guidelines and manuals have been issued for decentralized district planning by the District Planning Committee (DPC). The perspective plans and annual plans developed at village Panchayat, Block Panchayat, and District Panchayat are integrated with urban development plans at the DPC level to prepare the Integrated District Plan (IDP). The roles of Panchayats and Urban Local Bodies have been indicated in Schedule 11 and Schedule 12 in the 73rd and 74th CAA.

Schedule 11th contains 29 subjects as per 243 G which come under the power of Gram Panchayat. Panchayat needs to prepare its plan with the assistance of the planning team and standing committee members, which needs the approval of Gram Sabha. Panchayat Samiti would prepare its own plan, and inter-Panchayat plans, and integrate the GP plans into its own plan. The Zilla Parishad needs to prepare its own plan and integrate the Block plans and Urban Local Body Plans to prepare the Integrated District Plan which gets the approval of the Zilla Parishad/Panchayat.

Rural Development

A population living in a low-density and small settlements is called rural. According to Late (1975), "development is a process of improving standards of the masses of the low-income population residing in rural areas making the process of rural development self-sustaining". Hence rural development is a strategy to improve the economic and social life of the rural poor, including small and marginal farmers, tenants, and landless. It is an improvement in the living standards of the masses of low-income population residing in rural areas and making the process self-sustaining. According to Agarwal (1989), rural development is a strategy designed to improve the economic and social life of the rural poor. There are many such definitions given by different scholars on rural development. However, it is commonly understood that rural development is an integrated and multi-sectoral activity for all-out development in rural areas.

Concept of Rural Development

Rural development is a development process to make rural people self-reliant through various programs and projects. It is the joint responsibility of the government and local people. The benefits of development extend to the poorest seeking a livelihood in rural areas. The group includes small-scale farmers, tenants, and the landless.

As a strategy, it is designed to improve the economic and social well-being of the rural poor and it is multidisciplinary in nature, representing an interaction of agricultural, social, behavioral, engineering, and management sciences. It is a process of developing and utilizing natural and human resources, technologies, infrastructural facilities, institutions and organizations, and government policies and programs for speedy economic growth in rural areas.

In the words of Chambers (1983), rural development is a strategy to enable a specific group of people, poor rural women and men, to gain for themselves and their children more of what they want and need. It involves helping the poorest among those who seek a livelihood in the rural areas to demand and control more of the benefits of rural development. The group includes small-scale farmers, tenants and landless.

Need and Importance of Rural Development

The following broad activities are required for the sustainable development of rural people through RD programs.

Holistic development: RD is needed for economic and socio-cultural development of rural people and their living standards and self-esteem of rural youth, children, and women.

Community empowerment: RD is required for human resource development in terms of psychology, skill, knowledge, attitude, and other abilities.

Provision of basic needs: Provision of rural infrastructure and minimum facilities like drinking water, education, transport, electricity, and communication.

Institutional development: Development of institutions like Panchayat, Cooperatives, banks, post offices, and credit institutions.

Economic development and alleviation of poverty: Provision of financial and technological support for agriculture, industry, animal husbandry, forestry development, marketing, and natural resource management.

Recreation and employment: RD is required for the employment generation of rural mass and the development of recreation and entertainment facilities.

Basic Elements of Rural Development

Rural development has assumed significance under decentralized planning as their planning and implementation has been transferred to Panchayati Raj Institutions (PRIs). The development of rural areas and improvement in the quality of life of the rural people is the major concern of the RD Department.

There are at least three basic elements of rural development. The first element is the provision of basic needs for life sustenance. These basic necessities include food, clothes, shelter, healthcare, and security. The second element of development is self-respect, dignity, or honor. The third element is freedom which refers to political or ideological freedom and freedom from social servitude.

Rural Development Programs in India

Rural Development Programs have been started in pre-independence and post-independent eras with different objectives.

Pre-independence experiments: Various location-specific rural development programs were experimented in India during the pre-independence era. Christian Settlement (1860–1920, Sri Niketan experiment (1921), Gandhian experiments (1915–1942), Gurgaon experiment (1920), Martandam experiment (1921), Vedchhi project (1924), Rasulia experiment (1934), and Firka Development Scheme of Madras (1946) were experimented in India during pre-independence period which are indicated below.

Name of program	Initiator	Location	Objectives
Christian Settlement (1860-1920)	Christian Missionaries	Missionary areas	Economic and social uplift of their converts by organizing them into village peasant settlements, employing them in factories/workshops, and training them in certain simple trades to enable them to earn their livelihoods.
Sri Niketan experiment (1921)	by Shri Rabindra Nath Tagore	Kalingram Pargana of West Bengal	All round transformation of villages through selected activities.
Gandhian experiments (1915-1942)	Mahatma Gandhiji	Sabarmati Ashram, Champaran, and Savagram in Maharashtra	Hand spinning, Khadi movement, training for rural upliftment, rural sanitation, and women's welfare, and mutual help through SHG.
Gurgaon experiment (1920)	The Deputy Commissioner of Gurgaon District of Punjab, Mr. F. L. Brayne	Gurgaon district of Punjab	Improvement of agriculture, women's education, health and sanitation.
Martandam experiment (1921)	Mrs. Spencer Hatch and Dr. D. S. Hatch of the Young Men's Christian Association	Martandam in South Travancore of Kerala	Self-help, advisory for the introduction of improved seeds, pesticides, and tools, demonstrations, and cooperative societies.
Vedchhi project (1924)	Jugatram Dave	Vedchhi in Surat district of Gujarat	Improvement of Adivasis through the Khadi movement and education.

Rasulia experiment (1934)	Hildar Cashimose, Ranjit Chetsingh	Hoshangabad in Madhya Pradesh	Educational improvement, farming, well-digging, and cement sunk work. It emphasized the value of group work and community organization.
Firka development scheme of Madras (1946)	Firka Development officer and Committee	Firka centers in Madras State	To address the different rural problems through certain short-term as well long-term plans.

Post-independence efforts: With the learning experiences from the pre-independence experiments myriad of RD programs have been implemented and are still continuing in India.

In Rural Development, the critical areas identified were health, education, drinking water, housing, and roads. During the Ninth Plan period, several anti-poverty Programs have been restructured to enhance the efficiency of the Programs for providing increased benefits to the rural poor. Self-Employment Programs were revamped by merging the Integrated Rural Development Program (IRDP), the Development of Women and Children in Rural Areas (DWCRA), the Supply of Improved Tool Kits to Rural Artisans (SITRA), the Training of Rural Youth for Self-Employment (TRYSEM), the Ganga Kalyan Yojana (GKY) and the Million Wells Scheme (MWS) into a holistic self-employment scheme called Swarna Jayanti Gram Swarozgar Yojana (SGSY).

The major RD programs which are being implemented by the Ministry of Rural Development are the Mahatma Gandhi National Rural Employment Guarantee Act (MGNREGA) for providing wage employment, National Rural Livelihoods Mission (NRLM) for Self-employment and skill development, Pradhan Mantri Awas Yojana Grameen (PMAY-G) for providing housing to BPL households, Pradhan Mantri Gram Sadak Yojana (PMGSY) for construction of quality roads up to habitations, National Social Assistance Program (NSAP) for social pension, Shyama Prasad Mukherjee RURBAN Mission, and Integrated Watershed Management Program (IWMP) for improving the productivity of the land.

The RD Programs can be Categorized as Follows

i) **Rural Poverty Alleviation & Employment**

   Area Development Program: Drought Prone Area Development Program (DPAP) in 1973, Command Area Development Program (CADP) in 1974-75, Hill Areas Development Program (HADP) in 1975, Desert Development Program (DDP) in 1975, Drought Prone Areas

Program (DPAP) and Integrated Wasteland Development Program (IWDP).

ii) **Integrated Development Program:** Small Farmers Development Agency (SFDA)/Marginal Farmers and Agricultural Labourers Development Agencies (MFALDA) in 1971, 20 Point Program in 1975 and Integrated Rural Development Program (IRDP) in 1980.

iii) **Wage employment Programs:** National Rural Employment Program (NREP) in 1980, Rural Landless Employment Guarantee Program (RLEGP) in 1983, Food for Work Program (FFWP) in 1977, Jawahar Rozgar Yojana (JRY) in 1989, Employment Assurance Scheme (EAS) in 1993-94, Sampoorna Grameen Rozgar Yojana (SGRY) in 2001, The National Rural Employment Guarantee Scheme in 2004, and Development of Women and Children in Rural Areas (DWCRA) in 1982-1983.

iv) **Self-Employment Program:** Training of Rural Youth for Self-Employment (TRYSEM) in 1979, Support for Training and Employment (STEP), Swarnajayanthi Gram Swarozgar Yojana (SGSY) in 1999, and Sampoorna Grameen Rozgar Yojana (SGRY) in 2001.

v) **Infrastructure Development Programs:** Indira Awaas Yojana in 1985, Pradhan Mantri Gramodaya Yojana (PMGY) (Rural Shelter Component) for Rural Housing, Pradhan Mantri Gram Sadak Yojana (PMGSY) in 2000-01 for rural connectivity, Integrated Rural Sanitation and Water Supply Project with DANIDA Assistance, Central Rural Sanitation Program (CRSP) for rural sanitation, Sarva Shiksha Abhiyan/Samagra Shiksha Abhiyan (SSA) for rural education, The Accelerated Rural Water Supply Program (ARWSP) in 1972-73, Rajiv Gandhi National Drinking Water Mission in 2005, Swajaldhara for rural drinking water supply, Rajiv Gandhi Grameen Vidyutikaran Yojana (RGGVY) in 2005 for rural electrification, Public Distribution System PDS/TPDS) and Annapurna for food security and Accredited Social Health Activist (ASHA), Integrated Child Development Service (ICDS), National Drinking Water Mission (NDWM), National Rural Health Mission (NRHM), Swachh Bharat Mission (Grameen) for rural health & sanitation.

Brief Description of Major RD Programs in India

Etawah Project (1948): The project was launched in 64 villages in Etwah district in Uttar Pradesh by an American engineer, Albert Mayor. The project aimed at developing self-confidence and co-operation among the rural people.

The interventions were the use of an improved variety of seeds, chemical fertilizers, irrigation systems, improved implements, plant protection measures, soil conservation, improvement of animal husbandry, and co-operatives. The project was constrained by inadequate personnel and training but generated some new ideas.

Nilkhori Project (1948): After the partition of Pakistan, 7000 displaced persons settled in the new Town of Nilkhori. Shri. S.K. Day wanted the refugees to be actively engaged in constructive programs assuring them the right to live, the right to work for a living, and the right to receive what is earned. The project was started by Sri S.K. Dey organized dairy, poultry, tannery, printing press, engineering workshops, etc. on co-operative basis. The project was not successful due to a lack of experience and immature idealism.

The Bhoodan Movement (1951): Acharya Vinoba Bhave launched this program in 1951 for the distribution of gifted land to the poor by many state governments. It was a comprehensive movement for Gramdan as a logical consequence of the Bhoodan. The distribution of land took an unusual long time on account of the different legal hurdles and impediments. Most of the lands gifted were of poor quality and barren.

The Intensive Agricultural District Program (IADP) and Intensive Agriculture Area Program (IAAP): IADP was launched in 1961 for the application of modem technology and modem management practices in selected districts. In 1964, the IADP was extended to other districts of all States in the name of the Intensive Agriculture Area program (IAAP). Basically IAAP followed IADP but its staffing pattern was somewhat reduced.

High Yielding Variety Program (1966): The Yielding Variety program introduced in 1966 aimed at securing a breakthrough in agricultural production by means of effective measures of intensification through modern agricultural inputs and technology. The programs made significant improvements in agricultural production but received a setback due to severe droughts during the period 1966-1967. The program withstood the test of time and brought the "Green Revolution" in India, making her almost self-sufficient in food grain production.

Small Farmers Development Agency (SFDA) and Marginal Farmers and Agricultural Labourers Development Agency (MFAL): The government started 46 pilot projects for SFDA and 47 for MFAL in 1970-1971. SFDA aims at assisting persons specially identified from target groups and raising their income level, by providing them with agricultural inputs, including credit for the adoption of improved farm technology and to diversify their activities. MFALA was providing institutional credit facilities for undertaking various economic activities like horticulture, animal husbandry, and dairying, etc. Evaluation studies indicated that the benefits given to agricultural laborers were less and credit was not adequate to these small and marginalized farmers. Extension service was also inadequate.

Tribal Area Development Program (TADP), 1972: The program was launched in 1972 at six selected tribal areas for the development of tribal and backward areas. The program was an additional program of the normal flow of investment in the tribal areas on the ongoing schemes. The core of economic development pursued by these projects comprised agricultural development including horticulture, land reclamation, land development, soil conservation, and cattle development.

Watershed Development Program (WDP): Based on the recommendations of the Hanumant Rai Committee watershed development programs like the Drought Prone Area Program (DPAP), Integrated Wasteland Development Program (IWDP), National Watershed Development Program for Dryland Areas (NWDPRA), and Integrated Watershed Management Program (IWMP) along with DDP were implemented.

Draught Prone Area Program (DPAP) was introduced in 947 blocks of 149 districts of 13 states in 1973-1974. The areas with less rainfall, drought prone with no such facilities for irrigation were covered in this program. Both the Central and State governments spent in equal ratios for the implementation of the program. The broad objectives were the development and management of land and water resources, planting of trees, development of animal husbandry, crop management, and development of subsidiary industries. From 1995 to 1995 the program was implemented as watershed projects in 4975 watersheds of the country.

Desert Development Program (DDP) was implemented in 227 Blocks of 36 districts in Rajasthan, Haryana, Gujarat, Andhra Pradesh, Karnataka, Himachal Pradesh and Jammu & Kashmir) in 1977-78. This was an integrated development program of desert areas with interventions like planting of trees, water resource development, electrification of villages, and livestock improvement with the objective of increasing production, income of people, and creating employment opportunities.

The watershed development projects were implemented for soil and water conservation, plantation and fodder development, and crop improvement along with income-generating activities. The program thrust was on the development of proper technology for water management to minimize the adverse effect of draught upon animals and crops, integrate efforts to prevent the process of the desert, encourage balanced ecology, and improve the socio-economic conditions of people by resource development and technology. The watershed projects were managed by the Ministry of RD & Ministry of Agriculture. Subsequently, a common guideline was framed for the implementation of watersheds. The responsibility for the selection of micro-watersheds of about 500 ha and the implementation was given to the District Rural Development Authority (DRDA). The Project Implementation Agency (both government and non-government) selected who were responsible for implementation with the assistance of the Watershed Development Team (WDT), Watershed Association (WA), Watershed Committee (WC), and SHGs. Currently, all the watershed programs are implemented under IWMP which has been merged with Pradhan Mantri Krishi Sinchai Yojana (PMKSY).

Command Area Development Program (CADA): CADA was launched in 1974 for the improvement of irrigation efficiency in canal-irrigated areas. The interventions were land leveling, the construction of field channels, the introduction of the rotational water supply system, and the popularization of integrated soil, crop, and water management practices. Due to the rapid rise in the cost of all items such as labor, material, equipment, spares, land, etc., the cost of projects had automatically increased, and it was found difficult to complete the projects. There was political interference in the construction of field channels in the command area. Still, the program enhanced the irrigation efficiency in command areas.

Hill Area Development Program (HADP): HADP was launched in 1973-1974 to ensure planned Integrated Development of Hill Areas. It aimed at socio- economic growth with eco-restoration, eco-preservation, and eco-development in hill areas. Tribal Development projects were launched in 8 districts of AP, Bihar, MP, and Orissa in the last year of the Fourth Plan while Hill Area Development Program (HADP) was launched in two districts of UP and Manipur. In the Fifth Year Plan, one more district of UP was included.

The interventions were better utilization of water and land for agriculture and horticulture, improvement of land, water, and electricity, provision of the facility of housing for the landless farmers, minimization of traditional planting of trees and shifting cultivation, encouraging allied activities like dairy, goatery, piggery, poultry, apiary, sericulture, and sanction of loan.

Training for Rural Youth and Self-Employment (TRYSEM): The program was started in 1979 as a multi-pronged strategy to address rural unemployment and hence poverty. The selected beneficiaries were given skill development training of their choice with Government support for acquiring tools, equipment, and raw materials to set up small industries. There were some constraints on the selection of beneficiaries and sanction of loans by financing institutions.

Integrated Rural Development Program (IRDP): Integrated Rural Development Program is a very major program of poverty alleviation and rural development. The target groups were small and marginal farmers, agricultural laborers and rural artisans. The approaches followed for SFDA, MFAL, DPAP, and CAD programs were proposed to be utilized according to their relevance in particular areas to generate employment opportunities and increase production. The main objectives are creating assets, employment, increased income removal of poverty, and minimizing in equality. The programs were mainly Agriculture Development, including efficient utilization of land and water resources with scientific technology, animal husbandry as a subsidiary occupation and programs under the fishery, social forestry, and village industries. The IRDP concept was launched in 1978-1979 in 2300 blocks in the country and it was extended to all blocks in the country with effect from October 2, 1980. Simultaneously, SFDAs were merged with IRDP.

Special provision was made for SC/ST families with 50% reservation, for women with 40% reservation, and 3% for physically handicapped. Expenditure on subsidy was shared by the Center and the State on a 50:50 basis. The subsidy was limited to 25% of the project for small farmers, 33.3% for marginal farmers, agricultural laborers, and rural artisans and 50% for SCs, STs, and physically handicapped. Any rural poor family whose income was less than Rs. 3500/- per annum in 1979-1980 was the beneficiary, which has been enhanced to Rs. 11,000/- per annum now.

Indira Awas Yojana (IAY): Indira Awaas Yojana (IAY) was launched during 1985-96 as a sub-scheme of RLEGP and as a sub-scheme of Jawahar Rozgar Yojana (JRY) subsequently in 1989 as a housing scheme in rural areas. From the year 1993-94 the scope of IAY was extended to cover Non Scheduled Castes/Scheduled Tribes below the poverty line families in the rural areas. IAY has now been de-linked from JRY and has been made an independent scheme with effect from January 1st, 1996.

The scheme is primarily to help the construction of dwelling units by members of Scheduled Castes/Schedule Tribes, freed bonded laborers, and also non-SC/ST rural poor below the poverty line by providing them with grant-in-aid. A maximum of 40% of the total IAY allocation during a financial year can be utilized for the construction of dwelling units for non-SC/ST BPL categories. From 1995-96, the IAY benefits have been extended to widows or next-of-kin of defense personnel and Para military forces killed in action irrespective of the income criteria subject to the condition that (i) they reside in rural areas; (ii) they have not been covered under any other scheme of shelter rehabilitation; and (iii) they are houseless or in need of shelter or shelter up-gradation. Benefits have also been extended to ex-servicemen and retired members of the paramilitary forces as long as they fulfill the normal eligibility conditions of the Indira Awaas Yojana and have not been covered under any other shelter rehabilitation scheme. The priority in the matter of allotment of houses to the ex-serviceman and paramilitary forces and their dependents will be out of 40% of the houses set apart for allotment among the non-SC/ST categories of beneficiaries. A fund to the tune of 3% is earmarked for the benefit of disabled persons below the poverty line. This reservation of 3% under IAY for disabled persons below the poverty line would be a horizontal reservation i.e., disabled persons belonging to sections like SCs, STs, and Others would fall in their respective categories. The beneficiaries are identified by the Panchayat on the basis of the allocation made by District Rural Development Agencies (DRDAs) / Zilla Parishads. Gram Sabha will select the beneficiaries from the list of eligible households according to IAY guidelines and as per priorities fixed, restricting this number to the target allotted.

National Social Assistance Program (NSAP): NSAP was introduced in 1995 to give assistance in the form of pension to elderly people, widows, and disabled people under the BPL category. Indira Gandhi National Old Age Pension Scheme (IGNOAPS), Indira Gandhi National Widow Pension Scheme (IGNWPS), Indira Gandhi National Disability Pension Scheme (IGNDPS), National Family Benefit Scheme (NFBS) and Annapurna are included in the scheme. While the Central Pension is uniform across the states and UTs, the pensions given by the states vary from state to state as per their resource capacity and number of beneficiaries.

The pension under the IGNOAPS is Rs 200 per month for persons above 60 years up to 79 from the Center while persons over 80 will get Rs 500 per month. Under IGNWPS, BPL widows aged 40-59 years are given a monthly pension of Rs. 200/- while IGNDPS provides assistance to BPL persons aged 18-59 years with severe and multiple disabilities with a monthly pension of Rs. 200. Under National Family Benefit Scheme (NFBS a BPL household is entitled to lump sum amount of money on the death of primary breadwinner aged between 18 and 64 years. The amount of assistance is Rs. 20.000. Under Annapurna, 10 kg of food grains per month are provided free of cost to those senior citizens who, though eligible, have remained uncovered under NOAPS. The states are supplementing with variable pensions under this scheme in addition to central assistance.

Million-Well Project (1998-1989): MWP was launched to provide the facility of irrigation from the bigger wells for such people as small and marginal farmers living under the poverty line, scheduled caste, and tribal people and laborers.

This was introduced under the 'Employment Guarantee Program' and the 'National Rural Employment Program' for the landless people. In April 1989, joining both employment programs, a new program was formed which was named 'Jawahar Rozgar Yojana'. The responsibility for the implementation of this project is of the District Rural Development Agency on the district level and of Gram Panchayat at the village level.

Swarnjayanti Gram Swarozgar Yojana (SGSY): SGSY was launched in 1999 as a Centrally Sponsored Scheme for self-employment of poor. The program was a restructured program of IRDP, DWCRA, SITRA, and GKY, besides MWS. The poor families above the BPL (Swarozgaris) were provided income-generating assets through a mix of Bank Credit and Governmental Subsidy. The program covers all aspects of self-employment of the rural poor viz. organization of the poor into Self Help Groups (SHGs) and their capacity building, training, selection of key activities and planning of activity clusters, infrastructure build-up, technology, and marketing support. Each Block concentrates on 4-5 Key Activities based on local resources, occupational skills of the people, and availability of markets. Special safeguards have been provided to vulnerable sections, by way of reserving 50% benefits for SCs/STs, 40% for women, and 3% for disabled persons. Subsidy under the SGSY to individuals is uniform at 30% of the Project Cost subject to a maximum of Rs. 7500/-. In respect of SCs/STs, the subsidy is 50% of the Project Cost, subject to a maximum of Rs. 10,000/-. For groups of Swarozgaris, the subsidy is 50% of the cost of the scheme, subject to a ceiling of Rs. 1.25 lakh. SGSY is being implemented by the District Rural Development Agencies (DRDAs), The Swarnjayanti Gram Swarozgar Yojana (SGSY) is financed on a 75:25 cost-sharing basis between Centre and State Governments.

Mahatma Gandhi National Rural Employment Guarantee Act (MGNREGA): Started in 2005 as the National Rural Employment Guarantee Act (NREGA), it guarantees 100 days of employment in a financial year to any rural household whose adult members are willing to do unskilled manual work in a financial year to any rural household whose adult members are willing to do unskilled manual work.

The adult members of rural households willing to do unskilled manual work may apply for registration either orally or in writing to the concerned Gram Panchayat which issues job cards to the eligible workers. A written application is to be made by any of the registered members of the household to the GP or Block Office stating the time period for which the work is sought and the GP is to acknowledge the receipt of the written application against which the guarantee to provide employment within 15 days operates. The job card holders are provided manual unskilled work for the construction of durable assets in rural areas in lieu of payment of minimum wages in cash of DBT. In order to create durable assets a shelf of projects is prepared at the Panchayat level and approved at the Panchayat Samiti level. The prioritized works are agriculture-related, rural connectivity and irrigation etc. The labor and material component is usually 60:40.

The Gram Panchayat and the Program Officers shall ensure that every applicant shall be provided unskilled labor work as per provisions within 15 days of application for work or from the date from which work was sought whichever is later. In case employment is not provided within 15 days, the State (as per the Act) will pay an unemployment allowance to the beneficiary. Wages are to be paid as per the state-wise Government of India (GOI) notified MGNREGA wages according to piece rate as per the Schedule of Rates (SORs) without any gender bias. GP gets priority for NREGS works as a minimum of 50% of the allotted amount to a district is given to the GPs whereas a maximum of 30% at the Panchayat Samiti level and 20% at the ZP level is spent.

The Act calls upon every State Government to make a Scheme (National Rural Employment Guarantee Scheme, (NREGS) to ensure the livelihood security of the poor through the creation of durable assets that would lead to sustainable development or rural areas. Under NREGS durable assets are created leading to improved water security, soil conservation, higher land productivity, strengthening drought proofing and flood control, and rural connectivity. MGNREGA envisions 100 days of guaranteed employment as a matter of right to every rural household and strengthens decentralized, participatory planning and execution through the convergence of various anti-poverty and livelihood initiatives. Sampoorna Gramin Rojgar Yojana (SGRY) and the National Food for Work program (NFWP) were merged with NREGS.

Pradhan Mantri Grama Sadak Yojana (PMGSY): In order to provide rural connectivity up to habitations, the government has launched the Pradhan Mantri Gram Sadak Yojana on December 25th, 2000 with all-weather access to unconnected habitations. PMGSY is a 100% Centrally Sponsored Scheme and 50% of the Cess on High Speed Diesel (HSD) is earmarked for this Program. All-weather Road (with necessary culverts and cross-drainage structures, which are operable throughout the year) is constructed under the program. Unconnected habitations with a population of 1000 persons and above are covered in three years (2000-2003) while all unconnected habitations with a population of 500 persons and above by the end of the Tenth Plan Period (2007). It was 250 per habitation in Schedule V areas. The PMGSY permits the up-gradation (to prescribed standards) of the existing roads in those districts where all the eligible habitations of the designated population size have been provided all-weather road connectivity. In upgradation works, priority is given to the Rural Core Network.

Role of Extension in Rural Development

The structure of the Indian economy remained predominantly agricultural and development plans aimed at increasing food-grain production to meet the growing demand of people. The technological developments focused on the use of modern inputs irrigation, and farm mechanization to improve crop productivity of land. In a mixed farming system agriculture animal husbandry and fisheries are integrated with agriculture. Production of a mere 50 million tons of food grains has increased to about 330 million tons in 2022-23. The productivity of dry-land regions is still low and food security has been affected by climate change. The smallholders have low farm income which is lower than the consumption expenditure. It has become imperative that the extension agencies operating in the rural sector should work for the development of all sections of the population and transfer improved technology in agriculture and allied activities.

Agricultural Extension is mostly concerned with communicating the technology of scientific agriculture to the farmers to increase their production, income, and standard of living. It focuses on adoption of the new technology and improved practices in various sub-sectors like crop production, animal husbandry, fodder production, sericulture, beekeeping, horticulture, etc.

Agricultural extension workers are trained on technologies for agricultural production and post-harvest measures including farm management and community organization. However, the jobs associated with "rural extension" include micro-enterprise development, non-formal literacy education, family planning, nutrition, health, and other rural, non-agricultural areas. Therefore, a multi-sectoral extension covering both agricultural and non-agricultural extension would be required for rural development. Hence, it is desirable for the agricultural extension workers to analyze the felt and unfelt needs of the farming community in the rural sector and prepare extension plans for their holistic development. Motivating different farmers' groups could be done through proper planning of the adoption process within the entire community. Farmers should get direct attention to new and better farming techniques through the use of various extension tools. They should be motivated to take an interest in better technologies and should develop a desire to adopt the innovations. Once they are convinced, adoption will be easy.

The extension system aims at developing the individuals and the society as a whole. Therefore, the analysis of various groups in rural communities is most important to bring about the desired change in the attitude of the rural community toward accepting the new technological change in agriculture. Different farmers within a group may like to get advice from different leaders depending on their specific needs.

The modern extension process aims at working with rural people through informal education to achieve total community development covering several activities in the on-farm, off-farm, and non-farm sectors. Therefore, Agricultural extension is regarded as a special branch of rural extension dealing with several economic and social aspects of the farming community. These activities include efficient agricultural production, marketing, agro-processing, conservation of natural resources, and improving the standard of living of farm families by alternative income generation, development of leadership, and community development.

Role of Extension Worker

The role of the extension worker is very important in rural development. An ideal extension worker must attend to the following duties:

- Motivate rural people for the development of programs so that they themselves may become anxious to improve their standard of living and participate in the implementation of development programs.
- Identify local felt and unfelt needs and prepare development programs based on such needs.
- Create a self-belief in rural people to work on a self-help approach and make them understand that they can do most of the work through mutual cooperation.
- Educate the people to become self-dependent without waiting for outside assistance.
- Establish a close linkage between the research and farmers to address the problems of farmers through scientific solutions.
- Motivate the people to make full use of locally available resources.
- Organize the extension activities for social, economic, and educational development programs for all round development including education, health, transportation, electricity, water supply, etc.
- Should try to bring change in the mental attitude of people so that they can sacrifice their selfishness for the welfare of society and participate in the reconstruction of the country.

New Vision of Rural Development

The objectives of rural development are extended beyond agriculture and consider the development of off-farm and non-farm activities, village industries and micro-entrepreneurship, and physical infrastructure like roads, electricity, irrigation, etc. There is a need for improvement of human capital, social capital and natural capital. This requires a multi-pronged strategy and inclusive approach. It is a challenge to help poor farmers gain access to savings and micro-credits for agricultural production or micro-enterprise development. This requires strengthening the management and program development skills of extension workers, developing both the ongoing services and seeking the participation of private extension services including efforts of NGOs. The agricultural extension is required to perform the challenging task of food security, poverty alleviation, and enhancing farmer income.

Agricultural extension requires to be strengthened through a wider variety of institutional interventions than just public sector extension. Convergence strategy may also be required by involving different stakeholders. The importance of stakeholder participation and demand-driven community development projects are now well recognized. Governments may take a renewed and re-envisioned role in rural development through a network of extension institutes.

References

Agarwal, S. 1989. Pre-Post Rural Development. International Journal of Socio-Legal Analysis and Rural Development, 2(1), 97-106. Retrieved March 01, 2019 from http://osou.ac.in/eresources/PRE-POST-Independence-Rural-Development.pdf.

Chambers, R. 1983. Rural Development: Putting the Last First (Volume 198). Longman, London. Ministry of Panchayati Raj (1992); 73rd Constitution Amendment Act, 1992.

Suggested Reading

A.R. Desai. 2005. *Rural India in Transition*, 2nd ed. Mumbai: Popular Prakashan Mumbai.

Caroline, A. 2001 *Emerging Issues in Rural Development. An Issues Paper*. London: Overseas Development Institute.

Chauhan S.P. 1997. "50 years of poverty Alleviation programmes in India", Employment News, 30th August-5th September.

David A.M. Lea and D.P. Chaudhuri 1986. "Rural Development and the State" Published by Methuen & Co., New York.

Dwarakanath, H.D., 1995. "Rural Development Policy and Programme Implementation - An Evaluation." Social Action Journal, (1995) December.

Farrington, J.; I. Christoplos; A.D. Kidd; and M. Beckman. 2002. "Can Extension Contribute to Rural Poverty Reduction? Synthesis of a Six-country Study." ODI, Agricultural Research and Extension Network, paper no. 123. London.

Government of India 1996-97. Ministry of Rural Areas and Employment Information Brochure. New Delhi.

Gupta R.K., and Srivastava, S.P. 1989. "Action Sociology and Dynamics of Rural Development." New Delhi: Ajanta Publications, New Delhi.

Kartar, Singh. 1986. Rural Development: Principles, Policies & Management. New Delhi: Sage Publications.

Kothari, R.C. 1991. "Rural Development Vol. I Strategy for Rural Development". Manak Publications.

Maheshwari, S.R. 1995. Rural Development in India: A Public Policy Approach. New Delhi: Sage Publications.

Mathur, Y.B. 1985. "Rural Development in India 1885 to 1985." published by Hyderabad: National Institute of Rural Development, Hyderabad.

Mehta, S.R. 1984. "Rural Development Policies and Programmes:" A Sociological Perspective. New Delhi: Sage Publication, New Delhi. 31.

Padhye, Kishore Chandra 1986. "Rural Development in Modem India". B.R. Publication. New Delhi. Rao, B.S.S. 1992. 'Television for Rural Development" Concept Publishing Company, New Delhi.

Reddy, Venkata. 1988. "Rural Development in India: Poverty and Development". Delhi: Himalaya Publishing House, Delhi.

Vyas, V.S., Bhatt, A., and Shah S.M., (1985.) Decentralised Planning in India. New Delhi: Oxford and IBH Publishing Co.

World Bank. 1997. Rural Development: From Vision to Action. ESSD Studies and Monographs Series No. 12. Washington, DC. World Bank. 2002. Summary report of the World Bank/ USAID/Neuchatel International Workshop on "Extension and Rural Development" held at IFPRI. Washington, DC.

10

Participatory Rural Appraisal and Participatory Learning and Action

Initially, rural development followed a top-down approach based on the information collected through the traditional survey method. It was soon noticed that the perception of the development professionals from the "top" was at variance with the perception of the rural masses. Visits were made and extractive surveys were carried out with the help of structured formats and questionnaires. These surveys had several biases such as spatial, season, person, caste, religion, project, education, etc. These limitations led to a search for a new method for understanding rural situations. New methods and approaches such as Farming System Research (FSR), Exploratory Survey and Rapid Rural Appraisal (RRA) were attempted during the 1970s to overcome the deficiencies of conventional survey methods. Moving away from extractive survey questionnaires, two families of approaches and methods, referred to as Rapid Rural Appraisal and Participatory Rural Appraisal evolved in the 1980s and 1990s.

Origin of PRA

Participatory Rural Appraisal is '*an approach and methods for learning about rural life and conditions from, with and by rural people*'. It is defined *as a family of approaches and methods to enable rural people to share, enhance and analyse their knowledge of life and conditions, to plan and act.*

According to Robert Chambers (1992), there are five main trends that decorate the principal method of PRA: Activist Participatory Research, Agro-ecosystem Analysis, Social Anthropology, Field Research on Farming Systems, and Rapid Rural Appraisal.

a) **Activist participatory research**: As per the suggestion of Paulo Freire (1968), the poor can and have the possibility to analyze their own facts and conditions. Adult Participatory research has been associated with the adult education movement. The contribution of activist participatory research to PRA has been more through concept than methods.

- Poor people are creative and capable of making their own investigation, analysis, and planning.
- Outsiders have a role as convenors, catalysts, and facilitators.
- The weak need to be empowered.

b) **Agro-ecosystem analysis:** Based on systems and ecological thinking, agro-ecosystem analysis has combined the systems and system properties (productivity, stability, sustainability, and equitability) with many methods of analysis. The major contributions of agro-ecosystem analysis to RRA and PRA have been the transects, informal mapping, diagramming, and scoring & ranking.

c) **Social anthropology**: Social anthropology has been concerned more with applied anthropology and development anthropology. Social anthropologists helped development professionals generally better appreciate the richness and validity of rural people's knowledge. The insights of social anthropology have been the idea of field learning with flexibility, the value of participant observation and conversation, the importance of attitude, behavior, and the validity of indigenous technical knowledge.

d) **Field research on farming system**: Farming system research systematized the methods for investigating, understanding, and prescribing for farming system complexity. Many scientists focused on the participation of farmers in research. It was increasingly recognized that farmers could play a greater part in agricultural research. The contribution of farming system research is understanding the complexity, diversity & risk factors of farming systems; the knowledge, professionalism and rationality of small holders; the experimental mindset and behavior for their own analysis.

e) **Rapid rural appraisal (RRA):** In order to overcome biases in rural development tourism such as spatial, project, person, gender, seasonal, and diplomatic, etc. RRA evolved in the late 1970s and 1980s. The family of approaches and methods known as RRA gained increasing acceptance by the professionals. It was considered cost-effective in getting timely information. The RRA methods, when properly conducted, elicited a range and quality information and insights that were not possible in traditional surveys. The main reason for developing RRA was to find shortcuts in the search for relevant information on rural development issues in order to avoid costly and time-consuming research procedures. RRA is mostly carried out by a small team of researchers or trained professionals in one to three days in a kind of workshop. In RRA, the villagers were passive actors whereas the outsiders were active actors.

Participatory Rural/Relaxed Appraisal

Roberts Chambers (1992), the father of PRA suggested that the local communities can conduct their own analysis to plan and take action by using PRA. It is a bottom-up approach with the participation of the community to find out the problems, needs, and potential of the village. As a paradigm shift from a top-down approach to rural planning, PRA was developed in the 1990s to avoid the limitations and biases of extractive field surveys. In contrast to RRA, the PRA emphasizes on empowering rural people to make their own analyses and take decisions. In PRA, the local people own the results and the outsiders or experts own the results. Rather the outsiders work as facilitators in PRA exercise and help local people conduct their own analysis, plan and take action. In PRA, the outsiders learn from rural people (Table 10.1).

Principles

There are certain principles which are common with RRA and PRA and some additional principles have been emphasized for PRA. The common principles are outlined below:

Some Principles that are Shared by Both PRA and RRA

- **A reversal of learning**: It is a reversal of learning to learn from local people, directly, on the site, and face-to-face, gaining insight from their local physical, technical, and social knowledge. Learning from, with, and by local people through their symbols, criteria, categories and indicators is the role reversal.

Table 10.1: RRA and PRA compared

Particulars	RRA	PRA
Period of major development	Late 1970s and 1980s	Late 1980s and 1990s
Main user	Aid agencies, universities	NGOs, government field organizations
Key resource	Local people's knowledge	Local people's capabilities
Main innovation	Methods	Behavior
Predominant mode	Elective, extractive	Facilitative, participatory
Ideal objectives	Learning by outsiders	Empowerment of local people
Longer-term outcomes	Plans, projects, publications	Sustainable local actions and institutions.

- **Offsetting biases**: It is relaxed process, unimposing and learning from poor people and women their concerns and priorities. Through different perspectives, methods and tools, sources of information, people from different backgrounds, and places are engaged in relaxed learning.

- **Rapid and Progressive Learning**: With conscious exploration, flexible use of methods, opportunism, improvisation, iteration, and cross-checking learning becomes rapid and progressive.
- **Optimizing trade-offs**: Relating the cost of learning to the useful truth of information, the trade-offs between quantity, relevance, accuracy, and timeliness including the principles of optimal ignorance are achieved. It brings a trade-off between the cost of learning and the usefulness of information. It is better to be approximately right than precisely wrong.
- **Triangulation**: It provides for cross-checking of information from different sources for progressive learning. Cross-checking can be done from different methods, sources, and disciplines.
- **Seeking for diversity and differences**: Maximizing the diversity and richness of information may involve sampling in a non-statistical sense. It goes beyond cross-checking and deliberately looks for, notices, and investigates contradictions, anomalies, and differences. The aim is to seek variability rather than averages. It provides for learning from exceptions, dissenters, etc.

Additional principles emphasized in PRA

- **They do it**: This is expressed as "handing over the stick" with confidence that they can do it. The facilitators can initiate the process of participatory analysis and then allow the people to do themselves without any interference. Outsiders should believe that farmers should be able to draw maps, make models, evaluate, score, analyzes make a plan, and carry out the plan.
- **Personal responsibility**: PRA practitioners tend to take personal responsibility for what is done rather than relying on the authority of manuals or any rigid rules, in the spirit of the sentence, "Use your own best judgement at all times" (Peter, 1989).
- **Self-critical awareness**: The facilitators continuously and critically examine their own behavior. This includes embracing error-welcoming it as an opportunity to learn and have to reflect on and evaluate their experience.

  Sharing: PRA provides sharing of information and ideas between local people and facilitators and between different practitioners. The information and new ideas should be shared amongst farmers, facilitators, and between farmers and facilitators

Three pillars of PRA

The above PRA principles are mainly personal: to do with behavior, attitudes, and self-awareness. This contrast indicates the emphasis in PRA on how outsiders should interact with local people. In reality PRA is founded on three pillars: behavior/attitude, method, and sharing. The three pillars of PRA are associated with a range of attitudes and behaviors, methods, and types of sharing. These are illustrated in Figure 10.1.

PRA heavily relies on participation, flexibility, teamwork, and optimal ignorance. PRA data are mostly qualitative for which statistical analysis is not feasible. It is very important for the PRA practitioners to follow some basic principles, behavior, and attitudes which are displayed in Figure 10.2.

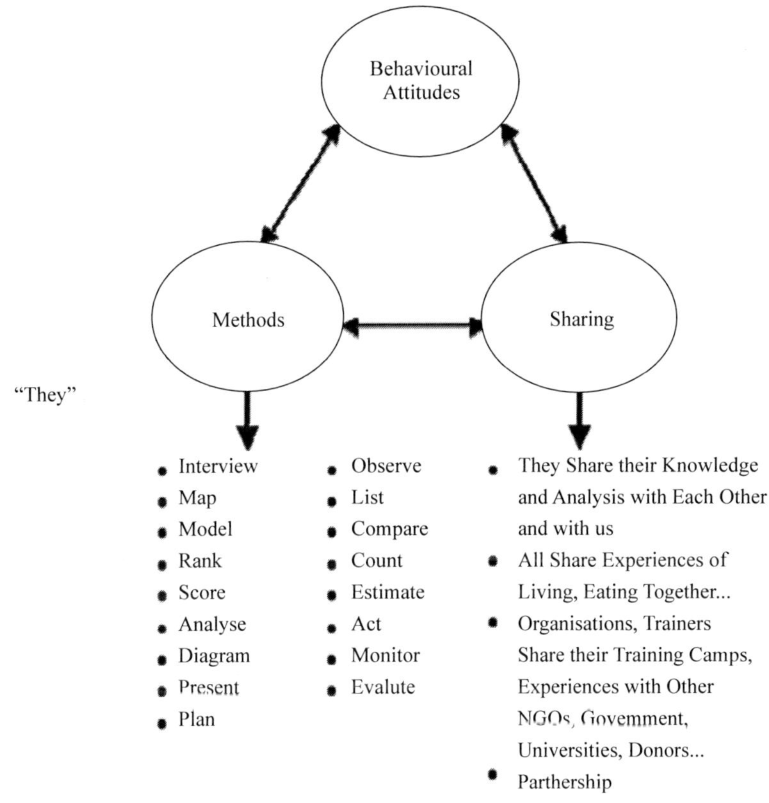

Figure 10.1: Three pillars of PRA (Chambers 1997).

Distinguishing Features of Participatory Rural Appraisal (PRA)

PRA is the result of a response to overcome the limitations of the 'top-down' approach to development, the pitfalls of the questionnaire survey method of inquiry, and the problem of biases. Thus there have been major shifts and reversals in PRA as shown in Figure 10.3.

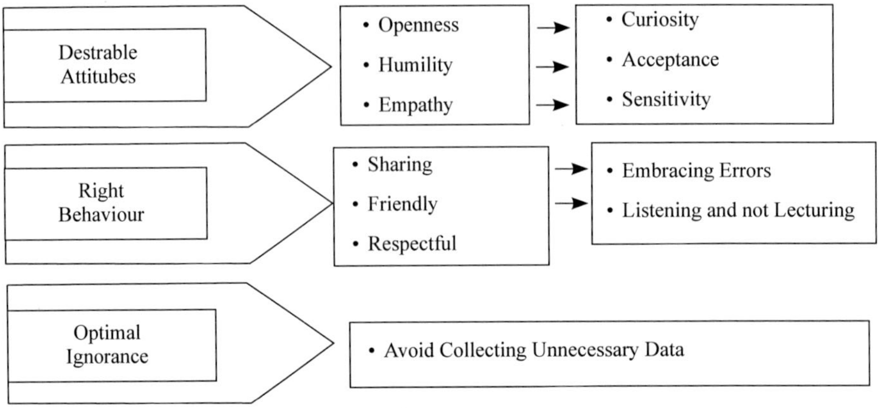

Figure 10.2: Principle, behavior, and attitude in PRA.

- **Closed to Open:** In questionnaires, the pre-set questions reflect the outsiders' understanding. In PRA, there is a shift from a closed system to an open system. The people determine what goes into a diagram, figure, or calendar. In a matrix, for instance, the local people do not just score but also determine the items to be judged as well as the criteria on which to judge. PRA encourages people to be involved and control the process following an open approach.
- **Measurement to Comparison:** Emphasis is given on absolute measurement in the traditional approach of inquiry. PRA gives emphasis on comparison as it is easier, faster, cheaper, and less sensitive than measurement. Trends and changes are captured easily and quickly even where baseline data are not available.
- **Individual to Group**: In both the top-down approach to development and the questionnaire method of inquiry, the basic element of the interface between outsiders and the people is the individual respondent. In PRA, it is group work, and methods like transect, maps, calendars, matrices, scoring, ranking, etc. are done by a group of people. It is true that semi-structured interviews take place with individuals, but the focus is on working with groups. The group interaction provides greater and more in-depth information and analysis in a much shorter time than the members could do individually in the same span of time in the non-participatory approaches.

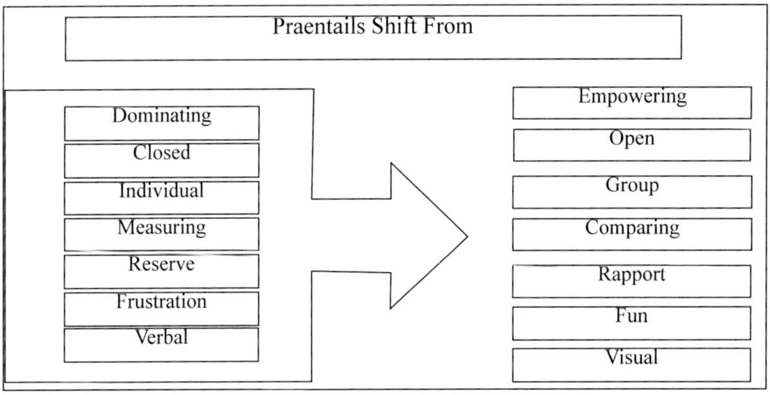

Figure 10.3: Shift in PRA.

- **Democracy of the Ground**: Another shift in PRA is from tables and paper to the ground. This has many advantages. The use of the ground means that more people can participate in parallel as equals. The use of the ground also has an anti-elite bias which allows the not-so-literate and not-so-articulate to participate actively along with the elite and the literate. The elite among the local people find working on the ground uncomfortable and, in certain areas, beneath their dignity. The problem with paper is that those who are not very familiar with the use of paper and pen feel inhibited. Making modifications to items and exercises are done on paper becomes difficult. There is a strong tendency among people to carry on with mistakes to avoid making the map all over again. This method is flexible and people can correct the errors easily.
- **Verbal to Visual:** The use of visuals in PRA has many advantages. The non-literate and not-so-articulate persons are able to use visuals as a medium in which to express their priorities and realities. In drawing or making the visuals, the inhibitions are lost and the participants express themselves more openly. The making of the visuals allows the local people to think through the process and come out with points that they are not consciously aware of but realize as the process goes on. Visualization also helps the participants to see and understand the interconnections between various issues, which in other modes of interaction, is missing. When social mapping is done, for instance, once the purpose is explained and the process initiated, the participants get involved in the process. More than one person can be involved in doing the visual at the same time but in a verbal mode only one person can speak at a time. It is important to note that though the use of visuals

and symbols is a characteristic of PRA, the verbal mode still plays an important role in it. The verbal mode also supplements the visualization process. Generally, the outputs and diagrams are interviewed. This helps the outside facilitators as well as the local people to clarify various aspects related to the issue under study.

- **Reserve to Rapport**: PRA lays an emphasis on rapport with the local people. Rapport building makes the people familiar with the facilitators and increases their involvement in the exercise.

General Process of PRA Implementation

There are three main stages in implementing PRA, namely preparation, implementation, and follow-up.

Preparation

The preparation activities cover the following:

a) **Training**: Training is implemented if the executive institution does not yet have experienced staff in implementing PRA. It would be preferable if the training involves community members of the village who will be involved in PRA. This will improve the convenience and accelerate the familiarization of the PRA team with the people during the implementation.

b) **Establishing the PRA Team**: Participants that had been trained are grouped in a PRA team consisting of people with various educational backgrounds and experience, men and women, from various levels of social economy, not only from the program planning organization but also representatives of the insiders (from the community). This team will unite their perception in order to maintain the unity of the team. This team will also have distributed work: some will become moderators, process observers, participant observers, documenters, and translators. If the scope of the PRA activities is quite large, a PRA sub-team may be formed from the existing PRA team.

c) **Determine the objective of the PRA**: The established team determines the objective of implementing PRA, whether it is a specific objective such as to identify the community needs, a monitoring program, and evaluation program; or the strategic objective that is to apply the principles of PRA in everyday work by the field officers continuously with a mission to educate and empower the people.

d) **Establishing the PRA activity design**: After the objectives have been determined, the team continues by formulating the general plan of PRA implementation consisting of the formulation of "activity designs" consisting of determining information needs, selecting and arranging the PRA techniques in order, determining the source of information, and creating a temporary schedule.

e) **Preliminary visit**: After the creation of the PRA activity design, the PRA team makes a preliminary visit to get permission to hold the activity, get acquainted with the people, explain the objectives, identify the participants, and collect secondary data. The preliminary visit is held long before the implementation of PRA activities. During this preliminary visit, the team also studies the secondary data. It would be even better if the secondary data are discussed with the community. These data are intended to create an early description of the implicated village. The data sources may be written documents from the village organizations, statistical data, village maps, records from monitoring programs from other institutions, and so on. Besides studying the secondary data, the team should also observe community members to determine the people who may help by becoming a supporting committee that will help in organizing the consumption and accommodation, determining the time and venue, preparing the equipment and the materials, and so on.

Implementation of PRA

After all the PRA preparation activities have been completed, the team makes a field trip to start the PRA activities. The activities to be conducted during the trip are:

a) **Discussion of the objective, target, and process of PRA:** This activity will commence with the re-discussion of the objectives, purposes, targets, and process of implementing the PRA techniques.

b) **Information gathering discussion**: After the community understands, discussions are held to gather information. All the information that will be recorded will certain/specific topics. Those topics are described, studied, analyzed, and concluded especially concerning the problems and potentials. If in the implementation the PRA team is divided into a number of sub-teams, there must be time to share/exchange the findings of each sub-team and formulate the subsequent plan for discussion.

c) **Recording the discussion results:** The results of the PRA activities are documented by recording in detail all the opinions from the people, not only on the related topic but also others that are considered important.

d) **Presenting the discussion results:** The results of the PRA are to be presented to the community members for discussion.

e) **Formulating the program plan:** The problems are to be prioritized and considered in developing the program alternative. The programs that are developed are expected to be able to solve the problems previously listed.

Follow-up Actions

The program plans that are made collectively with the community will be implemented. The PRA facilitating team returns to the field to assist the community and facilitate the implementation of those programs.

Some Guidelines for Conducting PRA

Before

- Meet the villagers with an open and frank mind. Build rapport and discus about the PRA exercise.
- Identify the villagers who are willing to share their experiences.
- Always begin the interview in a relaxed mood and set a climate for discussion.
- Select a suitable place for interview and sit on the same floor with the people.

During

- Listen carefully, show empathy, don't lecture, and be patient.
- Be polite, gentle, and accommodative and do not interrupt during discussion.
- Respect the villagers, and respect their social customs.
- Participate wholeheartedly, lead discussion with open ended questions, and allow time for reflection.
- Carefully avoid the persons trying to dominate in discussion.
- Resolve the discussion around important issues and give equal chance to different groups.
- Use six helpers: what, when, where, who, why, and how.
- Try to find different villagers for different tasks.

After

- Thank the participants after the interviews.
- Sit down with the members, record all information collected and make a presentation for information of all and their approval.

Tools and Techniques used in RRA/PRA

The tools and techniques used in RRA and PRA are almost same. These tools and techniques can be used according to the situation or need.

The tools and techniques used in PRA can be grouped as space related, time related and relation methods as indicated in Table 10.2 and Figure 10.4.

Table 10.2: Common PRA Tools in Use

Type	Tool	Purpose
Space Related	Social Map	An image of the habitation pattern of an area, according to people's perceptions
	Resource Map	A visual representation done by locals to illustrate the natural resources of a community
	Mobility Map	A visual analysis of the mobility patterns of the local people
	Transect Map	A cross-sectional view of an area used to explore the spatial dimensions of people's realities. Particularly useful in natural resources management
	Agro-ecology Map	A visual map showing agro-ecological situation of a village
	Technology Map	A map showing the technologies adopted in the village

Table 10.2: ***Continued....***

Type	Tool	Purpose
Time Related	Timeline	An aggregate of various landmark events in their community as perceived by the local people
	Trend Analysis	A 'before and after' type of analysis that identifies the changes that have taken place across certain time milestones
	Historical transect	Indicate the trends that have taken place over a period of time.
	Seasonal Diagram	An image of people's livelihood conditions over a period of time, the communities' notion of an annual cycle
	Daily Activity Schedule	A representation of people's daily activities from the time they get up until the time they go to bed. Hours are generally the unit of analysis
	Dream Map	Projection of dream of people in a map

Rationale Tools	Cause-effect Diagram	A visual representation of people's perceptions of the causes and effects of a problem
	Flow Diagram	A picture of the analysis of people
	Venn Diagram	A visual analysis of the importance and closeness of various institutions to the participants as perceived by them
	Matrix Ranking, Pair-wise Ranking and Preferential Ranking	A preferences analysis method that allows for comparison of different elements against a range of criteria
	Force Field Analysis	A temporal analysis of changes in livelihoods on an individual, household, or community scale
	Pie- Diagram	A representation of the proportions of constituents in relation to the whole as allocated by the people can be useful to analyze income generation activities
	Livelihoods Analysis	A temporal analysis of changes in livelihoods on an individual, household, or community scale
	Well-being Analysis	An analysis and evaluation of people's well-being situation
	Cob-web Diagram	A visual tool used to show the relative performance of a number of individuals or groups on various dimensions
	Bio-Resource Flow	Inflow and outflow of commodities from the village
Discussion Methods	Focus Group Discussion	Focused discussion by the community on a major topic or problem
	Semi-Structured Interview	Interview of the key informants or community members by a group using a checklist.

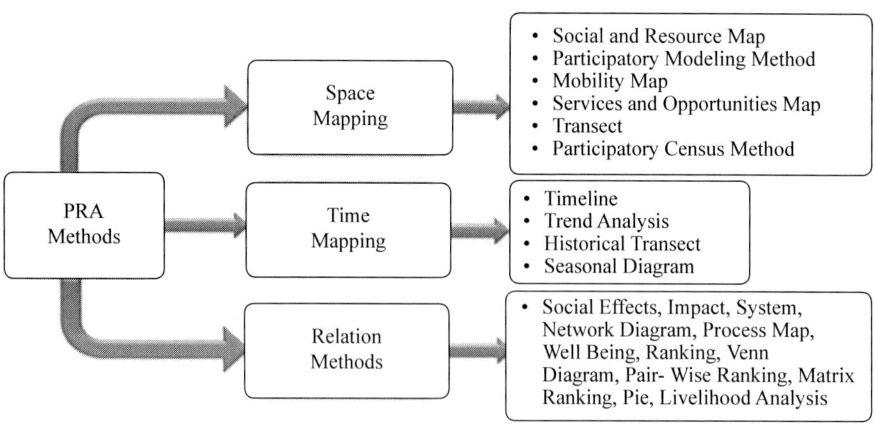

Figure 10.4: PRA methods.

Space related

a) Participatory Mapping and Modeling

Maps and diagrams are very important for planning activities. In rural development, maps are prepared to show the village layout, village infrastructures, houses, village resources, and living conditions of people for a greater understanding of the village. Mapping is the making of maps for the village that depict the conditions of the village area and its environment. The maps are prepared by the community as per their perception. Our experience shows that villagers produce much better maps on their own. These maps indicate the village infrastructures, natural resources, land ownership, settlement pattern, soil types, cropping pattern, etc. Such maps are drawn on the ground using locally available materials. The location of the mapping session should be freely accessible to all groups of the community (e.g. different castes, men and women, rich and poor, etc.). A future model can be drawn to discuss peoples' vision of the future. By collectively making the map for their area, the community will further recognize their village conditions, area boundaries, forest areas, gardens, settlement areas, etc. The main uses of participatory mapping are:

- Outsiders are able to get much information about the village and its resources.
- Outsiders get insight into the ways in which people think and prioritize their needs.
- Maps locate the situations of households and status of education, wealth, land holdings and livestock, etc.
- The resources are shown in the map indicating land, water, forest, and other natural resources.
- Shows the trends of both social aspects and economic aspects.
- Used for studying the village aspects and participatory assessment, planning, and monitoring the rural development programs at the micro-level.

1) Social Map

Social map of a village indicates the living sites of people, streets, infrastructures, and institutions that help to understand the situation of households and the availability of social facilities in a village with basic amenities. It serves as a baseline for planning, implementation, monitoring, and evaluation of village development activities (including selection of village organizing strategy) (Figure 10.5).

Objectives: The objectives of preparing a social map are to learn about social structures, living conditions of people, social institutions, and basic amenities available in the village and the infrastructural gaps for development planning.

Information collected

The representatives of the people are encouraged to draw the social map on a floor or open space to indicate the village boundary, roads, households (pucca and kuccha), availability of social institutions, and infrastructures in the village along with basic amenities. Some key questions can be posed during social mapping about a number of households by category, religious institutions, availability of electricity, drinking water, water bodies, drains, village problems, etc.

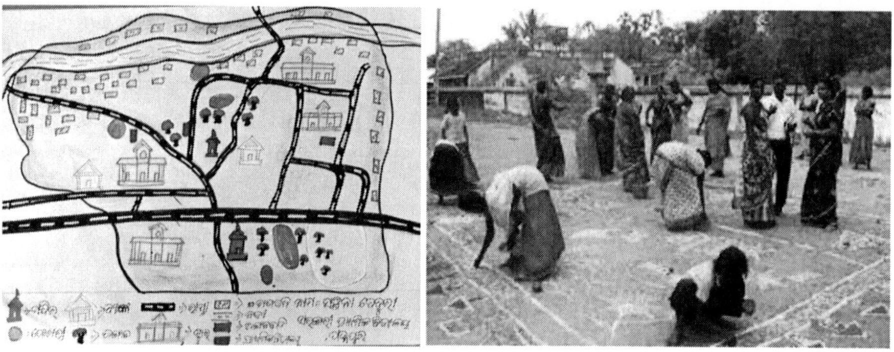

Figure 10.5: Village social map.

Facilitation

The willing villagers having some knowledge of drawing a map can be selected and a good site be selected for drawing a map using locally available materials like *rangoli* powder, wood ash, sticks, etc. to show different objects. The selected persons can use a village map available in the village to draw the map showing the approximate boundary, streets, roads, etc. They may be asked to locate all social and religious institutions in the village along with vulnerable households and women-headed households. The households may be located in different symbols to show rich and poor houses. The village ponds, open wells, hand pumps, schools, AWC and health sub-stations, market, Panchayat, drains, etc. are located on the map. Once the map is drawn, all people should see and make some corrections, if needed. After the map is agreed, a photograph can be taken up, and it may be converted to a paper sheet with different symbols.

Material required: Locally available material including wood ash, rangoli powder, stones, seeds, chalks, and cards of different colors.

Time required: The time required for the social map may vary from three to four hours depending upon the size of the village and the ability of the villagers.

II) Resource Map

Description: The village resource map is a tool that helps us learn about a community and its resource base (Figure 10.6). The village resources such as land, water, forest, grazing land, village commons, etc. greatly influence the livelihoods of people. A village resource map indicates these resources and the perception of people on these resources.

Objectives: The objective of preparing a resource map is to learn the villagers' perceptions of what natural resources are found in the community and how they are used. In order to gather relevant information on village resources, the facilitators need to put the following key questions.

Figure 10.6: Village resource map.

Key Questions:

- What resources are abundant?
- What resources are scarce?
- Does everyone have equal access to land?
- Do women have access to land?
- Do the poor have access to land?
- Who makes decision on land allocation?
- Where do people go to collect water?
- Who collects water?
- Where do people go to collect firewood?
- Who collects firewood?
- Where do people go graze livestock?

- What are the problems with these resources and what are the steps taken by the community for their preservation?

Facilitation

The villagers with a fair amount of knowledge in drawing a map are encouraged to draw the resource map side by side of the social map of the village in a suitable place which can be seen by a good number of villagers. After drawing the village boundary, the villagers are asked to locate different resources on the map by using different symbols with locally available material. Sometimes it is required to draw different maps by men and women at different places to understand their perceptions. Resource maps may include infrastructure (roads, houses, buildings, bridges, etc.); water sites and sources; agricultural lands (crop varieties and locations); soils, slopes, elevations; forest lands; grazing areas, common property resources, mining areas, etc. which are important for the community. Investigation may also be made by the facilitators about the problems of resource conservation and management in the village.

Material required: Locally available material including wood ash, rangoli powder, stones, seeds, chalks, and cards of different colors.

Time required: Time required for the social map may vary from three to four hours depending upon the extent of resources available, the area of the village, and the ability of the villagers.

III) Mobility Map

The mobility map is a PRA method used to explore the movement pattern of an individual, a group, or the community of a village (Figure 10.7). The focus is on where people go and for what. The map is drawn on a paper sheet or on a floor indicating the different institutions visited by the villagers and the distance of such institutions from the village. The map indicates the people's perception of movement patterns to these institutions and the average distance covered with frequencies.

Objectives

The objective of drawing a mobility map is to understand the mobility pattern of the villagers, evaluate the impact of certain interventions on their movement pattern, and plan for interventions and projects.

The mobility map provides information on:

- Distance (between the places of visit and their village)
- Mode of transport (bus or walk)

- Purpose of visit (weeding, sale purchase, wage labor, medical facility, etc.)
- Accessibility (in terms of size – the bigger the size the easier the access) In addition, it also specifies the places that they visit only in groups or with other member
- Frequency of visit

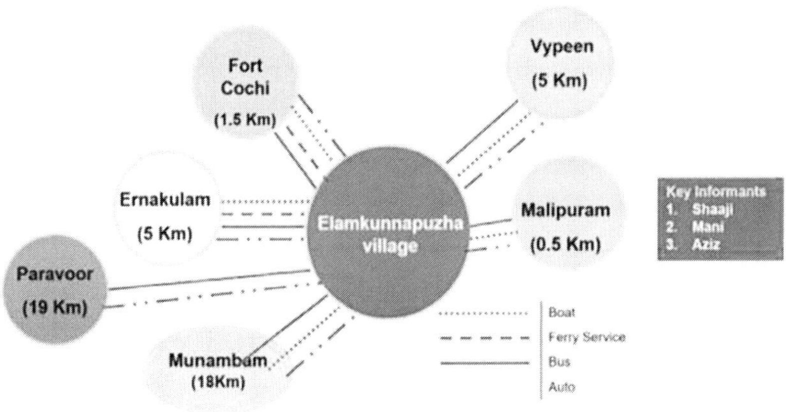

Figure 10.7: A typical mobility map of a village.

Process to be adopted

- Explain the purpose of the exercise and initiate a discussion on the places they visit and list down the places visited.
- select the person, group, or community who regularly moves outside of the village and whose mobility pattern you are interested in knowing;
- ask them to write the names of the places on small pieces of paper in bold letters or use symbols;
- draw a circle in the middle of a piece of paper or ground, representing the village/locality, and ask them to locate the places of visit by pieces of paper with the names of the places;
- ask them to link the cards representing the places visited with the circle depicting their locality by lines. The thickness of the lines could represent a particular feature, such as the frequency of the visits;
- ask them to follow a similar process, for all other places that they visit, one by one;

- ascertain the purpose of visiting the places, importance of the places visited, distance of the places, mode of transport, and frequency of visits, whether alone or with someone;
- ask them whether they would like to make any alterations once the diagram is ready; and
- request them to explain the map and their learning from it; and ask them to explain the diagram in detail.

Listen carefully to their discussion and take any necessary notes, copy the diagram onto paper with all the details, and triangulate the diagram and other details generated during discussions with others in the locality.

Material required: Locally available material including papers, colors, sketch pens, chalks, and cards of different colors can be used for the mobility map.

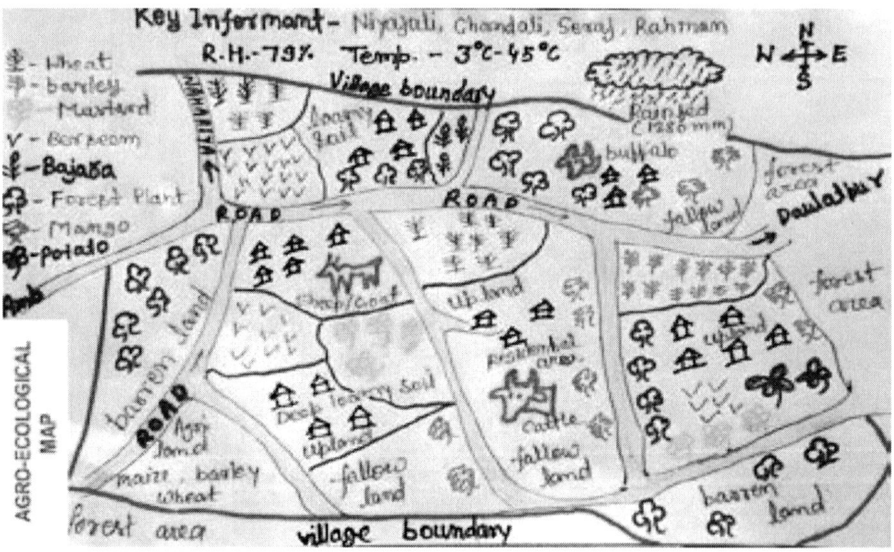

Figure 10.8: Agro-ecology map of a village.

Time required: Time required for the mobility map may vary from one to two hours.

IV) Agro-ecology map

Agro-ecology map will indicate the relationship between agriculture and the environment which includes average temperature, average rainfall, relative humidity, wind speed and direction, fragmentation of holdings, natural vegetation, drainage system, weeds, etc. The map indicates the biotic and abiotic components of the village eco-system (Figure 10.8).

Steps

- Explain the purpose of the map and encourage farmers to draw this map;
- identify major land marks;
- ask them to locate crop land, orchards, common land, etc. boundaries, river, hillocks, government land, forests, etc., up to the village boundary;
- depict crops, animals, natural resources like soil type, water resources (wells, river, channel, ponds, etc.), forest, Common Property Resources (CPR), use of locally available resources or whatever stakeholders observe during the walk; and
- write in local language along with English translation.

Questions to be asked for analysis:

Usually the following questions are asked to the participants during drawing of a agro-ecology map of the village.

- What natural circumstances you would like to probe for planning a research project?
- How do you identify these systems, subsystems and their boundaries?
- Who controls and uses Common Property Resources (CPR)?
- Do these resources depict the intersystem or intra system flow of energy?
- How can you use village transect information in preparing agro-ecological map?

The agro-ecology map of the village will help in deciding the potential of the village to take decision regarding various schemes suitable for the village in agriculture and allied sectors.

V) Technology map

Technology map is used to know different types of behavioral patterns of people towards technology adoption (Figure 10.9). The different types of technology behavior may be of adoption type, discontinuance, rejection and over adoption type, etc. Technology map comprises type and frequency of adoption of latest technologies through various agencies, related to agricultural research and development. The different technologies which have been adopted, discontinued and rejected in the village are to be ascertained. It helps scientists and extension workers identify the problems of the farmers through the feedback mechanisms.

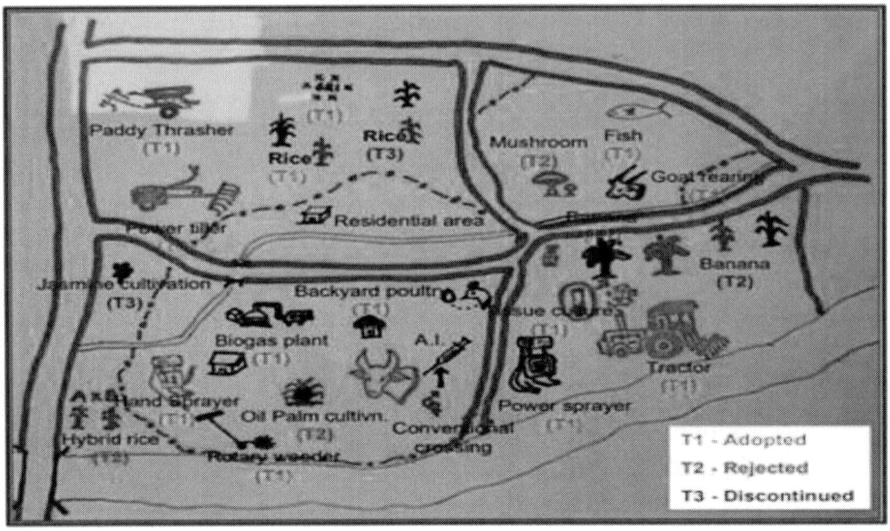

Figure 10.9: Technology map of a village.

Modeling

Modeling is advancement over mapping as it is three dimensional and shows greater details. In modeling, different models of houses, culverts, bridges, electric lines, etc. are prepared. This has been proved very useful in land use planning and watershed planning where problems, opportunities and treatments are indicated. Since models cannot be carried away like maps, they need to be photographed or copied in slides or on paper.

a) Village transect/Geographical Transect

Village transect or geographical transect is an exploratory walk through the area guided by the local people. It is used to explore the spatial dimensions of people's realities. It provides a cross-sectional representation of the different agro-ecological zones of the village indicating topography, land type, land usage, ownership, access, soil-type, soil fertility, vegetation, crops, problems, opportunities and solutions. It involves making a long walk inside the village along with key informant (KI) villagers and locating the various items that are found in the village like soil, crops, animals, problems, etc. and to identify the spatial differences. A geographical transect enables us to discuss on the above parameters with the farmers who accompany us. The discussions are real, situation specific, visual with the local community and outsiders. The interaction between the local people and outsiders helps establish a rapport and enhance participation.

Objectives

- Appraisal of natural resources in terms of current status, problems and potential;
- Verification of issues raised during other PRA exercises particularly during social mapping, natural resources mapping, etc.;
- Planning of various interventions and checking the relevance of the planned interventions; and
- Understanding the impact of interventions.

The process

The geographical transect could comprise of the following steps:

- Select a group of people, young and old, having some knowledge of the area with willingness to walk for the exercise across the village.
- Explain the purpose of transect to the people and involve them in the process of decision making regarding the transect path you should take to maximize the observing of details of the locality.
- Have a discussion and arrive at the parameters according to which you would like to collect data during the walk.
- If the situation on the field so demands, do not hesitate to make modifications. Also carry the list of parameters and preferably the resource map for the transect walk.

Observe the surroundings

The important things to be observed during transect are micro-environments, local technologies, problems and opportunities and non-agricultural livelihoods. However, it is preferable to take detailed notes with local terms, which can be translated afterwards. Ask questions to clarify things you are not clear about to the local people accompanying you. Listen carefully to what they say. While observing the cropping patterns, forests, etc. during transect several questions about productivity, technology used, problems and opportunities could be assessed which would be required for planning.

Types of Transect

There are several types of transects depending upon the purpose and resources available. The different transects used in PRA are village transects, classical/ straight transect, zig zag transect, looping transect and nullah transect. However, village transect or geographical transect is very common (Table 10.3 & 10.10).

Table 10.3: GT

Do	Don't
Arrange a willing group with field experience and plan the walk.	Don't take the transect in a haphazard manner.
Brief the group about type of information to be collected and start a low point and move towards a high point.	Don't be overawed by the exercise. Don't miss the opportunity to talk to passers-by.
Be curious and encourage discussion.	Try not to be bored yourself.
Start the exercise in morning and record observation in a map.	Don't walk too fast.
Make the exercise interesting for the participants and try to analyse the problems and opportunities on various spots.	Don't skip over important observations.

Items	Transect I	Transed II	Transed III
Land use	Crop Cultivation	Houses, Kitchen Garden	Plantion
Soil	Red Laterite,Ioamy	Red Ioamy	Red Laterite, Rainfed
Crops	Paddy	Onion, Chilles, Potato	Jamun
Trees	Arjuna, Sal, Jamun	Bamboo, Mango	Kendu, Sal, Eucalyptus, Sisso, Jamun
Animals	Desi Catties	Desi Caltles	Desi Cattles
Problems	Water Scarcity During Rabi Low Productivity	Un Employment in Lean Season	Soil Erosion
Opportu-nities	Constructing more Water Harvesting Structures, Introduction of High Yielding Short Duration Varieties, Agro Forestry and Introduction of Cross Bred Caltle.	Sericulture using For-est Trees, Introduced of Cross Bred Cattle	Gully Control, Afforesta-tion

Figure 10.10: Village transect map.

Time-related tools and techniques

i) Time line

Time line captures the chronology of events as recalled by local people from historical perspective. It is drawn as a sequential aggregate of past events which provides the historical landmarks of a community individual or institution. The important point to note here is that it is not history as much the events of the past as perceived and recalled by the people themselves. Time line can be used to explain the growth or decline of the technology or system of a particular activity. It gives an idea about the background of the village and evolution of specific activities like health, education, social amenities, agriculture, animal husbandry, fisheries, forestry, etc. The elderly people are the right persons who can give past information.

Objectives

- The objectives of a time line are to learn about the important past events of the village, to know the pace of development, to understand the changes with respect to issues on. Education, health, food security, gender relations, economic conditions, etc.

Process

The following steps are taken to prepare village timeline by the community.

- Identify some elderly persons in the village willing to talk about the history of the village invite.
- Explain them the purpose of the exercise. Initiate a discussion on the history of the village.
- Ask participants to note down the major events in brief on cards or on black board in bold letters. If the participants are unable to do so, ensure that one of the facilitators takes this role. Anyway, make a note of the key points and be willing to do this task yourself, if necessary.
- Read out the events and ask them whether they are happy with the order of if they would like to modify it.

Failure of memory, use of different time frames and calendar systems may present a big obstacle for the participants in arriving at the exact years. The facilitator may have to use own improvement or best judgment to arrive at the years. The timeline can be prepared for specific subjects such as agriculture, education, health, village development, natural calamity, etc. Some key questions which can be helpful in this regard may include: (What is the situation in the past? What were the major events? What changes have taken place? What were the reasons for change?) The pointes emerged from the time line could be noted down on the broad sheet of paper for future use. The names of participants, facilitators, location details, dates, legends, etc. should be noted down. There is a need for triangulation with other elderly people of the village or with secondary data.

Material required

The material required for time line includes cards, chalks, and chart paper, bold markers of different colors, seeds and pebbles.

Time required

Around 1 ½ to 2 hours should suffice for time line analysis. Time line can also be used as a precursor to taking up other time related methods like historical transect, trend analysis, etc. More focused information can be collected using other time related methods and the community can analyses how changes have taken place over the years. In fact, irrespective of the PRA method you have planned to use, a quick time line on the topic of your interest by participants can be helpful. Many issues may go unnoticed if a time line is not done at the beginning.

A time-line of a village relating to agriculture is presented in Figure 10.11.

Year	Agricultural Events	Symbols
1970	Bi-cycle, Radio	 
1975	Use of fertilizer	
1980	Tractor	
1985	Electricity	
1986	Irrigation	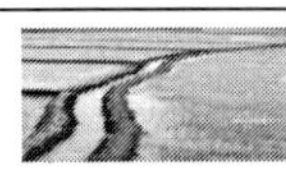
1990	Fishery	
1995	Watershed development	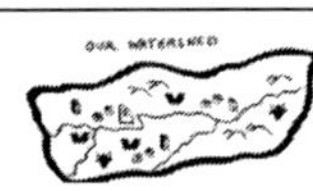
1996	Drought	
1998	Drip irrigation	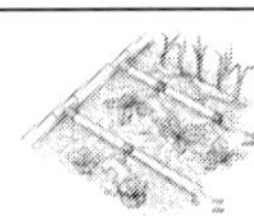
2000	Fruit plantation	
2005	Dairy farming	

Figure 10.11: Time line of agricultural events.

ii) Trend Analysis

Trend analysis captures the trends related to certain variables over different periods and indicates how things have changed from historical perspective. The local people have a good understanding of the present situation and the changes that have taken place over the years in their village. Trend analysis depicts quantitative changes over time in different aspects of village life, such as crop yields, population, livestock population, the number of trees, area under cultivation, rainfall, etc., which helps understand an increases or decreases in the variable under study over a period of time with possible causes.

Objectives

The objectives of this exercise are to learn about trend of development of many areas, resources, problems, impact of interventions for the change, and the present situation of these resources, etc. which will be required for future planning.

Information elicited

The following types of information are elicited from time trend analyses.

- Trend analysis of production and productivity;
- Price trend of commodities;
- Demographic trends;
- Trends in natural resources; and
- Population dynamics of livestock, etc.

It is also possible to carry out trend analysis across places, groups, individuals, community, etc., rather than over time alone. What has happened to particular aspects across various communities can also be called a trend, though it has been popularly used a time-related method.

Process to be adopted

- Select a group of local people who are interested in the exercise and explain them the purpose of the exercise.
- Initiate a discussion on the present situation and then move on to the aspects you are interested in pursuing, which sets the climate for trend analysis.
- Encourage the participants to depict the selected landmark years on cards preferably by symbols of visuals for representing various aspects.
- Ask participants to make the matrix on the ground, using chalk, and to represent from top to bottom the landmark years and from left to right various aspects like the density of trees, grass, wild animals, etc.

- Take up one of the aspects and ask the participants to depict the situation today in the relevant cell using symbols, visuals, seeds, sticks, sand, etc., as the case may be as per their choice.
- Move to the next time landmark and so on and after the completion of one aspect, move to the next one and follow the same process till the time all the cells are filled in.
- Once the diagram is ready, ask them whether they are satisfied with it or whether they would like to make any change.
- Find out whether they would like to add new aspects coming up during the process; and ask them to depict certain aspects that did not figure in their list but which you are interested in studying.
- Ask the participants to explain the diagram and encourage them to discuss their findings and reflect on them some key questions for the discussion could include: (Major trends and findings, Causes of the trends?, What can be done? Who can play a role in it? What can they do with a little assistance from outside?).

After the exercise is complete, copy the diagram onto a sheet of paper with details of the legend, the scoring system, the name of participants and facilitators, and the location and date. The findings can be triangulated from other sources.

The major landmark years are quite critical to the process of trend analysis. The landmark years should be decided based on two factors. First, they should be relevant to the topic of the study to capture the change in all their details and diversity. Second, they should be friendly to the participants so that they are able to relate to the years easily. If the landmark year is such that the participants are not really familiar with it, they would not be able to give a correct picture of the situation in the year. The events can be indicated by selecting fixed time intervals or identifying the years that may have implications for the topic of trend analysis you have selected.

Reflecting on trend analysis

In addition to the diagrams, the discussions, reflections, and analyses are very important. Interviewing the trend analysis provides an opportunity to enhance the understanding of the people's perception of trends of change over the year,clarify doubts about certain abrupt changes, and see how the problems can be solved in light of the aspirations of the local people.

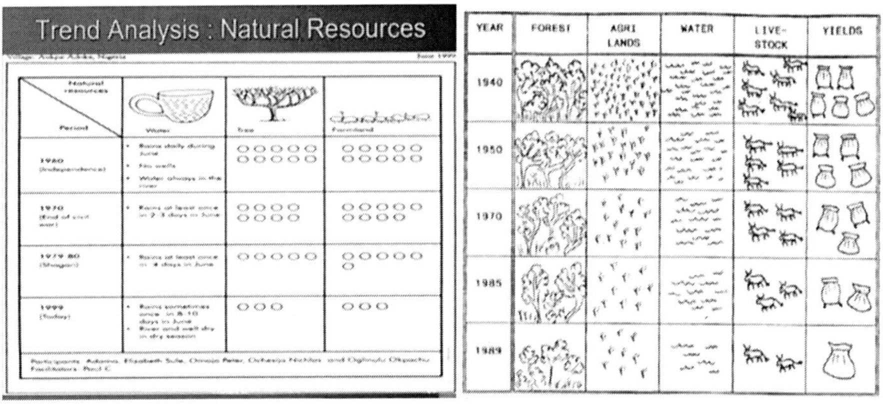

Figure 10.12: Trend analysis of natural resources

Material required

Cards, chalks, chart paper, and bold markers of different colors, seeds, and pebbles.

Time required

About 2-3 hours will be required depending on topics for discussion.

Scope for improvisation and complementarily with other methods

Trend analysis is a versatile PRA method that has been used in widely varied contexts and it complements other PRA methods quite well. An example of trend analysis of the natural resources of a village is shown in Figure 10.12.

iii) Historical Transect

It is a pictorial/graphic representation of the village area at different points in time, which shows evolutionary trends in land use, vegetation, erosion, population, etc. This exercise is done by interviewing older people and asking them to recap the landscape of a given area at different points in time. An example of a historical transect is shown in Figure 10.13. The map indicates the declining trend of forest and grazing land and cultivated land that indicates the declining population of cattle and crop production.

Figure 10.13: Historical transect.

vi) Seasonal Diagram (Calendar)

Seasonal calendar or seasonal diagram (Chamber 1992) by major season or by month shows the distribution of rain, crop activities, agricultural labor, diet, food consumption, type of sickness, prices, animal fodder, fuel, migration, expenditure, credit, etc. It is one of the popular PRA methods that has been used for temporal analysis across annual cycles, with months or seasons as the basic unit of analysis. It reflects the perceptions of the local people regarding seasonal variations on a wide range of items.

Objectives

The major strength of the seasonal analysis is that it depicts a range of items and their magnitude, which helps in understanding how these items are related to and influence one another. These relationships can be quite revealing. Seasonal diagram helps identify heavy workload periods, of relative ease, credit crunch, diseases, food security, wage availability, etc. This is an important PRA tool to analyze the important events and interactions, among them. For example, annual rainfall pattern can be considered for success of agricultural cropping, fodder availability, drinking water availability, labor availability, etc. The busy seasons, lean seasons, deficit periods, etc. could be ascertained from this calendar.

Procedure

The following suggested steps are recommended for making a seasonal diagram:

- Explain the objective of the exercise to the participants.

- Draw a grid with chalk on the floor with 13 columns and as many rows as the items you want to study.
- Keep the cards with names of the months and visuals or symbols in the top boxes in order, horizontally.
- Ask them to identify a unique characteristic of each month, one by one, that would remind them of the month.
- Start a discussion on the present month and then the work they have been doing during other months of the season.
- Encourage them to show the activities in different months by using pebbles, seeds or match sticks, etc.
- After completing one aspect or activity move to another, until all of them are similarly covered.
- Facilitate a discussion and analysis among the participants and others present to analyze the findings.
- Copy the diagram on a piece of paper with legends and details of the participants, facilitators, locality, and date.
- Thank the participants for their active participation and valuable Input
- Later triangulate to verify the findings with other key information to ensure that the information generated is correct.

The decision about the first month in the seasonal diagram should be left to the discussion of the people themselves. Experience shows that the rural and agricultural communities generally start the year from the month, which marks the beginning of a major agricultural season.

Seasonal diagrams done on the ground with chalks, seeds, powder, symbols, sticks, etc., have several advantages providing more space for people to participate, providing flexibility for change if desired, and enabling everyone to see what is happening and to contribute meaningfully.

Based on the literacy level of participants, different probing techniques will be applied. Experience has shown that people are quick to find something unique with which to represent the month. Symbols or diagrams used commonly include fruits and crops unique to the month, equipment, work, clothing, games, etc., unique to the month, unique items or articles associated with the festivals falling during the month, etc., seasonal aspects, e.g. rain snow, sun, etc. A sample seasonality calendar of a village in Odisha conducted by the author is shown in Figure 10.14.

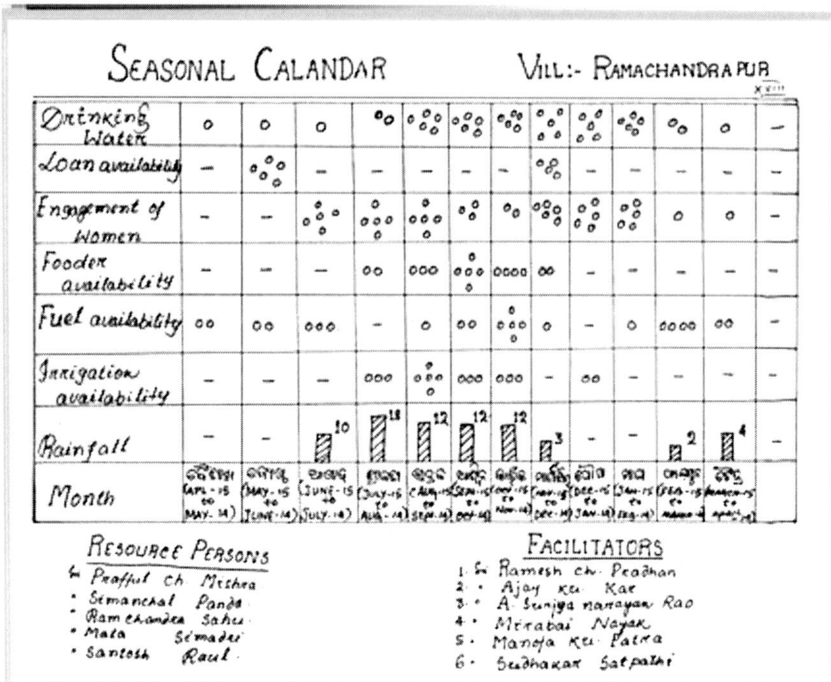

Figure 10.14: Seasonality calendar

Material required

Seeds, cards, marker pens, chalk of different colors, and other locally available materials like twigs, pebbles, etc., should suffice.

Time required

The time required for doing a seasonal diagram may vary depending on a host of factors, including the topic, interest of the participants, depth of information, and analysis aimed at. Two to three hours are sufficient.

v) Daily Routine

It is a discussion between the community on time utilization for each day for various groups of the community (rich, moderate, and poor).

Objectives

The profile of time utilization must be able to identify: (a) who does what, the work distribution in each house; (b) when the job should be done and how long will it take; (c) the difference in workload between various groups (rich, moderate, poor), and; (d) use of spare time in the community (for developing counseling activity).

4:00	Getting Up
5:00	Milking
6:00	Preparing Tea
7:00	Feeding the Animals
8:00	Cleaning the House
9:00	Cooking
10:00	Fetching Water
11:00	Bringing Wood
12:00	Mopping/ Cleaning/ Renewing the Floor
13:00	Washing Clothes & Taking Bath
14:00	Giving Fodder to Animals
15:00	Cooking
16:00	Animals Left for Grazing
17:00	Crushing/ Grinding Spices
18:00	Animals Brought Back
19:00	Feeding Animals
20:00	Milking
21:00	Cooking Dinner
22:00	sleeping

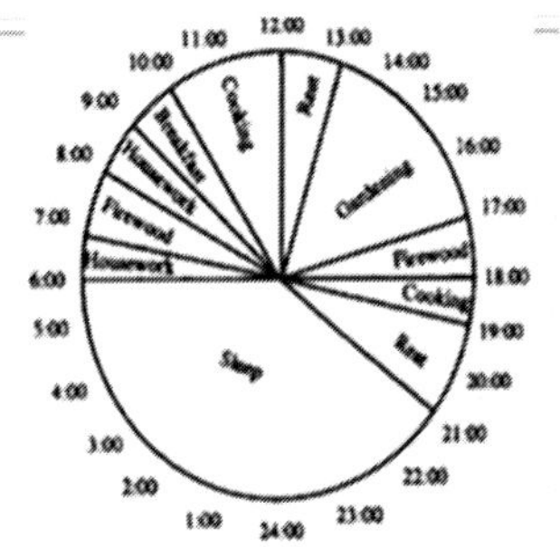

Figure 10.15: Daily routine for women.

Types of information that can be collected are daily activities of the community, time of execution, and work distribution in a family of each community group. This can be used to study the role of women in various activities. Figure 10.15 shows a model daily routine of women in rural areas.

vi) Dream Map

A dream map is used to understand the future aspirations of local people. What distinguishes a dream map from other types of maps is that it is an exercise where community members speculate their future situation in consideration of the present situation. The exercise shows a futuristic plan to solve their problems.

Applications

A dream map can be used for understanding the aspirations of village people, planning interventions in relation to areas where people desire change; and monitoring the progress against community goals.

Dream map prepared by different groups helps in arriving at the different perspectives of various groups about their future.

- A dream map of a village can be drawn by men, women, and children separately.
- A dream map of a school can be drawn by children, parents, and teachers separately.
- A dream map of a house can be drawn by various family members.

- A dream resource map of a village can be drawn by the landless and land-owning households.

Process

The suggested steps for dream mapping include:

- Select suitable persons from the community and explain them the purpose of drawing the map.
- After drawing the village map in an open space, the villagers will be asked to show and locate the aspirations by using visual symbols.
- After the dream map is made, showcase the two maps and initiate a discussion among the participants.

The main points for discussion may include:

- The problems of the present situation and their causes.
- the aspired state and ways of realizing it (including the actors who could help or hinder its realization.
- factors affecting the realization of their dream.
- their own role in the realization of their dreams.
- listing indicators against which the realization of their dream can be monitored.
- carefully ensure discussion, and take note of important points; and transfer the map to a broad sheet of paper with all details.

Advantages and Limitations

Dream mapping provides valuable insights into the present circumstances of a community, the aspirations of the people, and possible ways of improving the situation. If well-facilitated, dream maps can significantly improve the quality of people's planning. Sometimes it may unnecessarily raise the expectations of community members and once the high expectations are not fulfilled, the situation will be complicated.

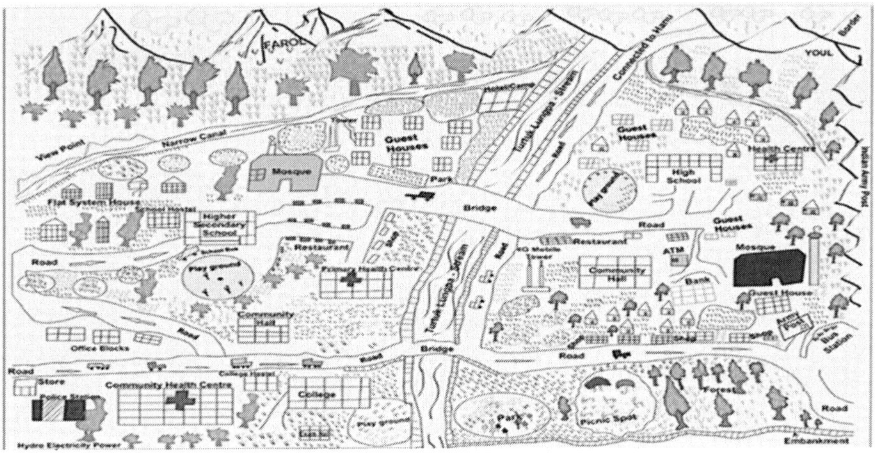

Figure 10.16: Village dream map.

A village dream map is presented in Figure 10.16.

vii) Occupational Analysis

Analysis of the occupation is a PRA activity consisting of discussion activities to recognize and analyse the conditions of the community life according to their occupation. Using this technique we can determine the types of occupations, work distribution, level of income and expenses of the community and the role of the men and women in the occupation sector. The information can be feedback and provide a description of the problems and needs of a community and the existing potentials and opportunities.

Objectives

- To identify the types of occupation in the community.
- To identify the potentials and problems from several community occupations.
- To reflect the previous efforts.
- To decide the feasible action for development.

Types of information that can be collected

- Occupations in the agriculture field such as dry crop plantation, farming, rice fields, garden and fishery.
- Non-agricultural occupations such as village industry, handicrafts, etc.
- Occupation in the field of services such as agricultural labor, barber, transportation, and village artisans.

Table 10.4: Example of an Occupational Analysis

Occupation	Number of households
Cultivation of crops	56
Horticultural activities	15
Animal husbandry	20
Poultry	10
Fishery	5
Forestry and NTFP collection	25
Agro-industries	5
Cottage industries	5
Wage labor	30
Village artisans	10
Petty business	5
Service	3
Others, specify	
Total households in village	**120**

Sources of information

The information can be collected from the community members, particularly people engaged in different occupation and from secondary sources (Table 10.4).

Relation Methods

Cause effect diagram (Problem tree and solution tree)

Cause effect diagram is a popular PRA method which falls under the family of flow and linkage diagram methods. It focuses on the causal factors of a phenomenon, activity, or problem, and the effects thereof. The cause effect diagram presents visually the causes, effects and their inter-linkage, which help in arriving at an in-depth understanding of a particular topic, and provides scope for analysis and subsequent action to be taken by the local people.

Objectives

Cause effect diagram has been used for the study and analysis of the problems of wide areas including food insecurity, poverty, illiteracy, the dropout rate of children from school, drinking or alcoholism, the status of women, migration, low productivity, drought, food insecurity, ill-health, etc. This helps solve the problem by suitable actions.

Process

- Decide on a topic for a cause effect diagram and invite a group of participants who are interested in the topic for the exercise.
- Introduce the topic to the participants. Explain the purpose of the exercise. Keep a sheet of paper with the topic or the subject of the cause diagram written in bold letters in front of the participants. Visual depiction is preferable.
- Ask the participants to focus on disusing the causes of the phenomenon. Note down the points in brief.
- Again read out the list of causes and effects. Ask them if they would like to make any modifications like deleting or adding new items onto the list.
- Handover the colored cards to the participants and ask them to depict as well as write down the causes on separate cards. Try to involve as many participants as possible in the process. Even if the participants are literate, visual depictions may provide others with an opportunity for getting involved.
- Show the completed cards and verify whether the participants can identify the symbols or diagrams.
- Repeat the same process for the effects using the other colored cards.
- The cards with causes can be placed in one side while the cards with possible effects can be put in the other side.
- Ask the participants to link the cards with chalk to show their linkages and connectivity.
- Listen to the discussions participants encourage in and note down the points.
- Once the diagram is ready, ask the participants to have a look, and make alterations if required. Note down the diagram on a sheet of paper with details.
- Ask questions on aspects about which you have doubts or want to get a more in-depth understanding.
- At the close of the exercise thank the participants for their cooperation and triangulate the diagram with others.

Flexibility in diagramming

The problem cause tree can be drawn on the ground with chalk which gives flexibility. The use of cards helps them see the linkage since cards can be physically moved and placed at various places to reveal the possible patterns.

It does not involve much effort and at the worst, the lines may have to be redrawn. For root cause analysis, they have to be asked a series of WHYs to arrive at the root cause. Once the root cause is found it will be easier to solve the problem.

Material required: The material required are cards of small sizes, seeds colors, markers, chalks, large-sized paper, etc.

Time required: The time required for cause and effect diagram depends on the topic, interest of the participant's scope for discussion, etc. The usual time required is approximately a minimum of two hours for an important topic.

Scope for improvisations and complementarily with other methods

A number of improvisations are possible with the cause effect diagram. You can give weightages or scores to the causes as well as to the effects. It gives an idea about the perceived importance of the causes as well as the effects. Ranking the causes and effects identified is another possibility. A simpler way out is to ask the participants to put the cards in a descending order so that the card on the top is the most important and the one at the bottom the least important. An example of cause and effect diagram have been shown in Figure 10.17.

viii) Flow diagram

A flow diagram is used to identify and depict the images of an activity, intervention or event. It takes into account the types of changes as perceived by the local people which helps identify impacts of certain events, whether planned, unplanned, negative, or positive.

In this method, local people describe cause-effect relationship, whether verbally or with the help of visuals and explain linkages among different factors. A flow diagram or flow chart can link problems with their perceived causes and help in arriving at possible solutions to that problems. This can improve the decision making power of the rural people. A flow diagram of hunger and poverty linked with food insecurity is presented in Figure 10.18.

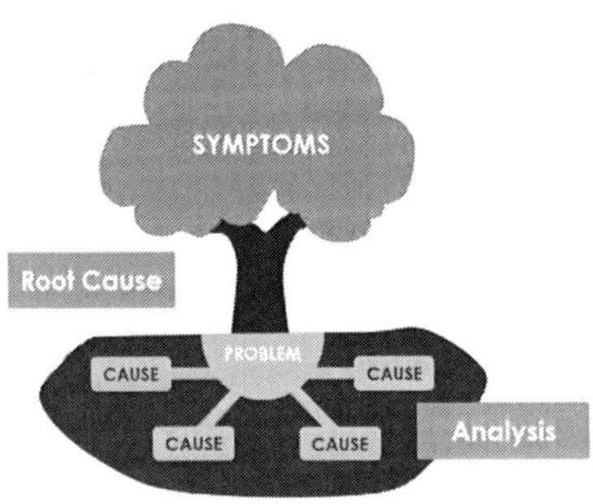

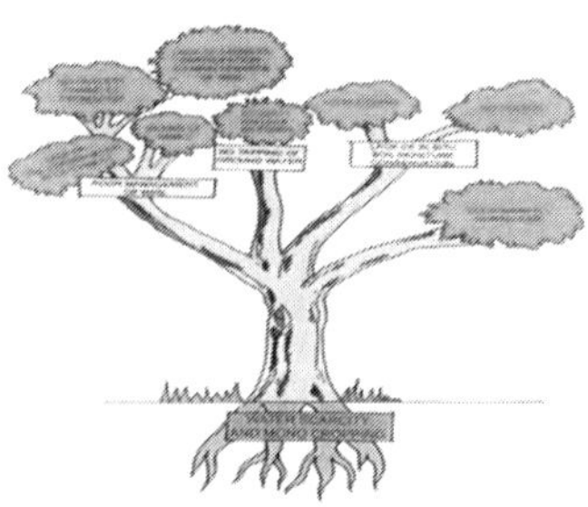

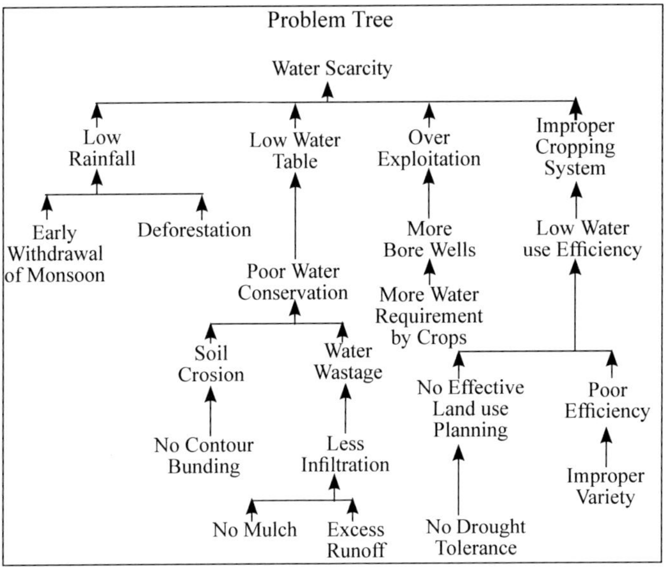

Figure 10.17: Problem cause tree.

ix) Venn Diagram (Chapati Diagram)

Venn diagram is used to study institutional relationship with the villagers and is also referred to as institutional diagram or *Chapati diagram*. This was named after English Logician John Venn in 1880s which was mostly used in mathematics. The method uses circles or paper discs of various sizes to represent institutions or individuals in order of importance. The important problems are shown in shape of bigger discs while less important problems are shown in smaller discs. The distance between circles represents the degree of influence or contact between institutions or individuals, which is a mental distance and not physical distance. Mutual interaction among the institutes can be shown as overlapping the discs.

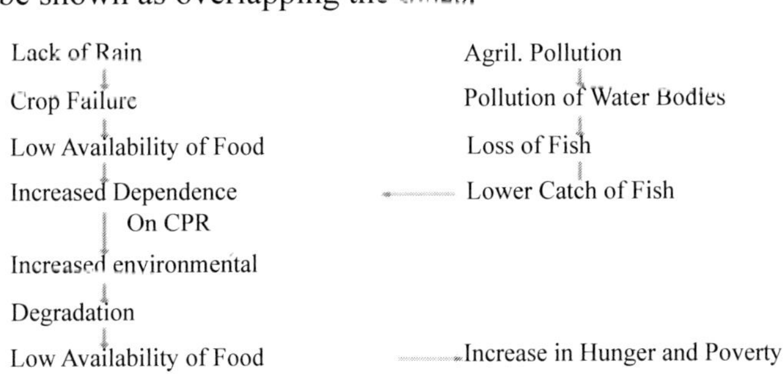

Figure 10.18: Flow diagram.

Objectives

The objectives of drawing a Venn diagram are to understand the importance of local institutions on the life of village people, their services available to the villagers and to identify the deficiencies with these institutions.

The relative importance of services and programs has also been studied using the Venn diagram which is particularly useful when you want to study and analyse (i) various institutions and individuals and their influence on the local people, (ii) various groups and individuals in the locality and their influence, and (iii) main actors in the community and their conspicuous and inconspicuous influence.

The Venn diagram elicits a wide array of information like (a) relative importance of various institutions in the village, b) relationship among them, (c) linkages among them, (d) weaknesses with respect to decision making process, (e) development of the village by institutions, (f) duplication of efforts among institutions, (g) gap identification between institutions, (h) objectives and felt needs of farmers, and (i) concentration of power within villages (Figure 10.19).

Figure 10.19: Venn diagram.

Process

The Venn diagram can be prepared by men and women separately to understand their perception on the village institutions.

- Select a convenient space such as open floor and select a small group of people and explain them the purpose and procedure.
- Ask them to list out the institutions in an around the village on which they are dependent and ask them to write their names on round paper

discs using big discs for important institutions and small discs for less important institutions.

- Once the cards are arranged in an order, ask them whether they agree or would like to make modifications.
- Encourage them to make changes, if they are interested.
- Ask them to locate the discs inside or outside the village in such a way that the bigger the circle, the higher that institutions or individual ranks on that variable.
- Paste on the circles the cards with names of institutions or individual.
- Draw a circle on the ground representing the community and circle in such a way that those high on the second variable, close together, while those low on the variable are kept representing the community.
- Ask them to locate the institutions at a distance place if they are not getting proper benefits from them, and put the useful institutions close to village.
- For clarification, ask them questions for their satisfaction or dissatisfaction.
- Copy the output onto a sheet of paper.
- Record the name of the village, participants, date, legends, what the size of the circle represents, and what the distance represents.

Thank the participants for their active involvement and time and triangulate the diagram and the major findings with others knowledgeable about the situation to ensure that your information is correct.

Problems in facilitation

In order to avoid the problems of facilitation, ask the participants to list the institutions which are relevant to them. Once the list has been mad ask them to put them in descending order based on each variable. Then ask them to assign paper circles of different sizes and so on and put those inside or outside the village. When two institutions have good interaction, they can be shown overlapping. In case one important institution (say, medical) has been shown on a big paper disc, but the villagers are not getting benefits due to absence of doctor, it can be placed at a longer distance from the village.

Material required

A brown paper sheet, small paper sheets for preparing discs, scissors, sketch pens, paste, etc. are required.

Time required

Usually, 2-3 hours will be required on the Venn diagram and the subsequent discussion.

x) Ranking & Scoring

Ranking is placing things on order relative to one another while scoring is giving things a number based on certain criteria. The methods of ranking and scoring reveal priorities and preferences of any intervention or problem. These methods are useful to learn from the people about their categories, criteria, choices, and priorities. Ranking methods include preference ranking, pair-wise ranking, matrix ranking, and well-being or wealth ranking.

a) Preference ranking

This exercise is done in local conditions to choose a topic preferably one which is related to the fieldwork. It enables the PRA team to examine the main problems or preferences of individuals and compare the result from different individuals. It allows the PRA team to determine quickly the main problems of preferences of individual villagers and enables the priorities of different individuals to be easily compared. It has logical steps to be followed:

- Decide the variables for ranking and arrange 5-6 persons to participate.
- Interact with the person and know his/her favored items in order of priority.
- Repeat this exercise with a good number of people.
- Tabulate the responses.

An example of preference ranking is shown in Table 10.5.

b) Pair-wise ranking

Pair-wise ranking helps villagers to set priorities (i.e. problems, needs, actions, etc.). This type of ranking can be undertaken with key informants or group of villagers that represents a good mixture of interests. For complicated issues, ranking can be undertaken using pair-wise ranking in order to determine the villagers' preferences.

The items for ranking are arranged and compared against each other. The reasons given by the respondent reveal the reason for the preferences of different persons/groups. Sometimes it is necessary to use different groups like young and old and men and women to know their perception.

Table 10.5: Constraints to Agricultural Profuction

Problems	Ranking by respondents					Total	Rank
	A	B	C	D	E		
Cost of inputs	1	2	1	1	2	7	6
Insect pest attack	4	3	4	3	4	18	4
Shortage of labor	5	5	3	5	1	19	3
Natural calamity	2	1	2	2	5	12	5
Moisture stress	3	4	5	6	3	21	2
Problems in marketing	6	6	6	4	6	28	1
6=Most Important 1= Least Important							

Procedure adopted

- Discuss with the villagers and decide the issues to be prioritized.
- Arrange a group of key informants from various categories and explain them the purpose and procedure for pair-wise ranking.
- Prepare a matrix on floor or on a broad sheet of paper and write down the names of variable vertically and horizontally in the same order.
- To get the villagers' preferences, facilitate the comparison of issues with one another. The first issue listed on the left side of the matrix will be compared with all the issues listed on the top. Repeat the process until all the issues have been covered.
- Note the number of times each problem was preferred. Make a summary of the preferences and rank them according to the score in ascending order.
- Give enough time to villagers to discuss and do not rush the exercise.
- At the end of the exercise, briefly discuss, analyse, and summarize the results together with villagers.

A model pair-wise ranking output of agricultural problems is presented in Figure 10.20.

Types	seeds	Tools	Insectlclde	Canals	Markets	Plpes	Transport	Trainig	Road	Score	Rank
Lack of Seeda		Tools	seeds	Canals	seeds	seeds	seeds	Trainig	seeds	5	3
Lack of Tools			Tools	Canals	Tools	Tools	Tools	Trainig	Tools	6	2
Lack of Insecticide				Canals	Insecticide	Plpes	Transport	Trainig	Insecticide	2	6
Canals					Canals	Canals	Canals	Trainig	Canals	7	1
Lack of Markets						Plpes	Markets	Markets	Markets	3	5
Lack of Pipes							Plpes	Trainig	Plpes	4	4
Transpot								Transport	Transport	3	5
Training									Trainig	6	2
Road										0	7

Figure 10.20: Pair-wise ranking of agricultural problems.

c) Matrix ranking

Matrix ranking is a better ranking technique which is based in certain criteria decided by the community. In matrix ranking criteria are used for the rows in a matrix and items are shown in columns. The items are ordered for each of the criteria and people are asked to put in piles of stones, seeds for scoring relative values. Ranking may be made in ten-point scoring or five-point scoring. Usually the variable getting most of the preferences is ranked at the top.

Purpose of Matrix Ranking

Matrix ranking has been used by researchers for various purposes like planning, institution ranking, livestock preferences, fodder preferences, constraints, and wild fodder, disease issues, animal losses, problem and solution ranking, success ranking household tasks and livestock. The criteria are chosen by the participants only. Matrices enable a range of different items to be assessed against selected criteria, preferences, attitudes and priorities of people toward a particular item and the reasons for the same can be known by matrix ranking. Selecting an appropriate crop for planting in a village by matrix ranking has been shown in Figure 10.21. Here, the villagers have preferred crop-B as their preferred crop.

d) Wealth Ranking

Well-being (or wealth) ranking is a relative and not an absolute assessment of people's wealth or well-being, and therefore provides insights on how the poor differentiate themselves amongst each other. People have their own perception about rich or poor in the society. The wealth ranking of households are done as per their criteria and perception.

Matrix ranking table on preference of Crops

CRITERIA	Crops			
	Crop A	Crop B	Crop C	Crop D
Farmer's/ Consumer's need/ demand/ taste	1	2	4	3
Availability of inputs / technology	1	3	2	3
Need for irrigation	4	2	3	1
Cost of production	1	3	2	4
Market availability	1	3	2	4
Sale price	3	4	2	1
Technical Feasibility	1	4	3	2
TOTAL	12	21	18	18

Score: 4= Best 3= Good 2= Fair 1= Worst

SCORES OF EIGHT TREES ACCORDING TO FIVE CRITERIA. GROUP-1A

PHOOLWARI VILLAGE MATRIX RANKING TREES. 1. MINIMUM 6. MAXIMUM

	TEAK	HALDU	KHAIR	KALLAM	SADAD	BAMBOO	MAHUA	EUCALYPTUS
TIMBER	☆☆☆☆☆☆	☆☆☆☆	☆☆☆☆☆	☆☆☆	☆☆☆	☆☆☆☆	☆☆	☆☆☆
FODDER	NIL	* * *	* *	* * *	* * *	* * *	NIL	NIL
FUEL WOOD	NIL	NIL	✿✿	✿✿	✿✿	NIL	✿	✿✿✿✿✿
AGRICULTURAL IMPLIMENTS	△▽△▽△▽	△	△▽△▽	NIL	NIL	NIL	△▽	△▽
MEDICINE	♠♠	♠♠	♠	NIL	NIL	NIL	NIL	NIL
OTHERS	NIL	NIL	NIL	NIL	NIL	♣♣♣♣	♣♣♣♣♣♣	NIL

Figure 10.21: Matrix ranking of preferred crop.

Applications

The exercise is helpful in:

- Identifying and classifying households or groups based on relative well-being in the areas of income, wealth, assets, status, etc.
- Studying inter-household and inter-group socio-economic disparities;
- Arriving at an understanding of people's criteria and indicators for wealth, well-being, development, etc.
- Assessing the impact of interventions or well-being programs for different groups/households and developing a baseline report for monitoring and evaluation of program.

Process

Well-being ranking can be done in different ways like card sorting method or social mapping method.

Card sorting method

In card soring method, the households are listed and the names of the heads of the households are written on small cards allotting one card for a household. Select the criteria as perceived by the villagers for deciding rich or poor.

The villagers are asked to rank the households by consensus as per wealth groups. While dealing with small villages with households ranging from 30 to 40, ask the participants to arrange the households in descending order of well-being. When dealing with larger villages, the above method may be too time-consuming. In such cases, ask the participants to sort out the household cards into representative categories of well-being. Ensure that the participants discuss amongst themselves and arrive at their own well-being categories. Find out the characteristics of each of the categories and note down the households as per different categories.

Social mapping method

This exercise is done after preparing the social map of the village. The villagers are asked to show the households in different colored cards as per categories of well-being. They are asked to decide the criteria for such ranking.

Advantages and limitations

Well-being ranking exercise is based on the criteria fixed by the villagers which proves to be incorrect in certain cases. In large localities or communities, and in urban or semi-urban areas, it is difficult to find participants who have an extensive understanding of different community groups. Wealth, income, assets, etc., are generally considered sensitive topics and are not discussed openly in many communities, therefore, well-being ranking can become a difficult exercise. There is also a strong tendency amongst people to present themselves as poorer than they really are in order to be eligible for benefits. It is necessary that the facilitator does not give any false promises to the villagers.

An example of wealth ranking is given below.

Wealth Ranking	Criteria
Rich HHs	Pucca house, big farmer with plough units, milch cow, government service, gold ornaments, two wheeler/four wheeler, lending money, double cropping in irrigated land, etc.
Average HHs	Kuccha /asbestos house, small farmer with ploughs, food security for 12 months, APL family, rain-fed farming, etc.
Poor HHs	Kuccha house, marginal/landless, wage labor, BPL, food security for 8-10 months, etc.

Criteria	Poorest of poor	Poor	Better of poor	Rich
House	Thatched cottage	Mud house	Asbestos house	Pucca house
Drought animal	nil	nil	1-2	2-3

Cattle	nil	1	2-3	3-5
Small ruminants	nil	1-2	2-4	nil
Assets	nil	nil	Bi-cycle, Radio	Motor Cycle, TV
Landed property	nil	0.2 ha	0.5 ha	2 ha
Main source of income	Wage	Wage & AH	Agriculture	Agriculture & Business
Food	6 months	10 months	12 months	Surplus for sale
Debt	Lending from Money lender	Lending from SHG	Lending from Banks	nil
Education	nil	nil	Primary	Secondary and above
HHs in village	10	20	50	10

v) Pie Diagram

A pie diagram is generally used as a tool to present information in a visual form that is based on factual data.

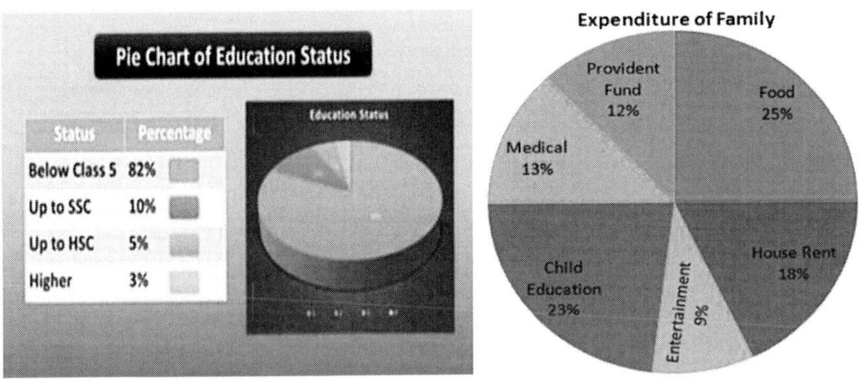

Figure 10.22: Pie chart showing family expenditure and education status.

Applications

Among other things, the pie diagram has been used to illustrate the population composition according to: caste, religion, tribe, occupation, literacy levels, etc. and the household details in terms of income and expenditure, etc.

Process

A group of villagers (both men and women) are asked to collect categorized information from the social map, etc. and then using that information to prepare a pie diagram. The 100 seed technique is a good method for understanding of the villagers. Suppose you are interested in finding out the proportion of expenditure involved with a child's education. The suggested process steps include:

- Explain the purpose of the exercise to the participants.
- Ask the participants (may be a family in this case) to list out all possible heads of expenditure and allot seeds out of 100 for each item. Count the number of seeds that have been put in each category. These numbers will provide you with the percentage of expenditure which can be easily converted into a pie diagram.
- Copy the output on a piece of paper, with details about the participants, facilitators, location, time, and make an index indicating symbols used.

The family expenditure on different items can be presented on percentage basis with the help of pie-chart (Figure 10.22).

xi) Livelihoods Analysis

Livelihood analysis is used to examine and depict the livelihoods of individuals or groups. The focus is on income, expenditure, food consumption, coping with crisis, livestock, agricultural production, employment related issues, etc.

Applications

The livelihood analysis provides detailed insights into the monthly or annual patterns of an individual's or a group's livelihood. The occupational and employment issues are explored at the community level, which provides a profile of the area. The individual livelihoods are examined, which tend to focus on income, expenditure, etc., as well as food consumption and distribution within the household

Process

A very commonly used approach is to start a livelihood analysis by making a social map, and then continue the process by identifying sources of income or expenditure for the households on the map. Different indicators can be used to decipher relative levels of income and expenditure for each household. A similar process can be conducted for other factors related to livelihood, including different types of employment, etc.

Advantages

In actuality, livelihood analysis is not one specific tool, but rather the adaptation of multiple tools to serve a specific purpose. The utilization of multiple tools also allows for a strong, comparative analysis between individuals, between men and women, and between different forms of livelihoods.

xii) Bio-Resource Flow

Bio-resource flow a village indicates the flow of inputs and outputs from and into the village. The bio –resource flow of a farm household also can be ascertained from the in-flow and out-flow of agricultural and livestock related products. The bio-resource flow diagram is useful in planning purpose. The bio-resource flow of a village is presented in Figure 10.23.

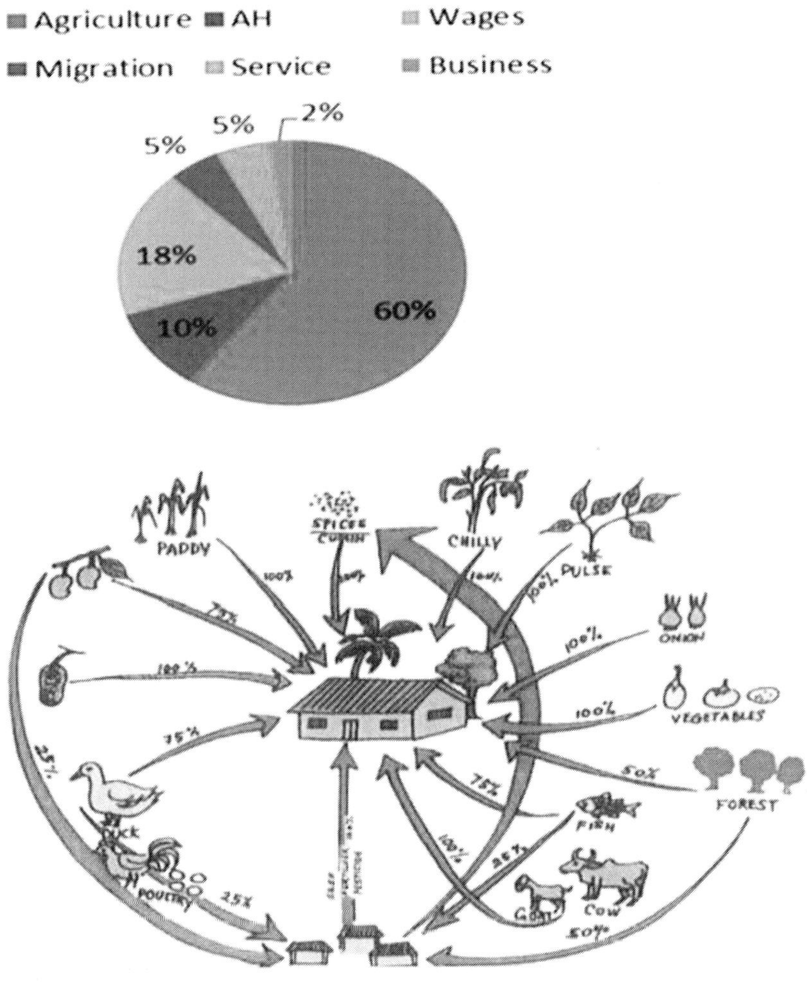

Figure 10.23: Bio-resource flow.

Force Field Analysis

Kurt Lewin (1940) originally used it in his work as a social psychologist. The idea behind Force Field Analysis is that situations are maintained by equilibrium between forces that drive change and others that resist change, as shown in Figure 10.24 below. The driving forces must be strengthened or the resisting forces weakened to make a desirable change.

Applications

Force field analysis is used for identifying the forces responsible for the present status of any aspect and analyze how the situation can be changed.

Process

- Write or draw the problem on a sheet of paper.
- Keep the sheet of paper in front of the participants ask them to visualize the problem, situation in a state of temporary equilibrium by two sets of opposing forces.
- Ask them to identify the forces for a change and that for a resistance and mark those on cards.
- Ask them to arrange the cards in two sides of the problem as positive or negative.
- Copy the diagram on to a piece of paper.

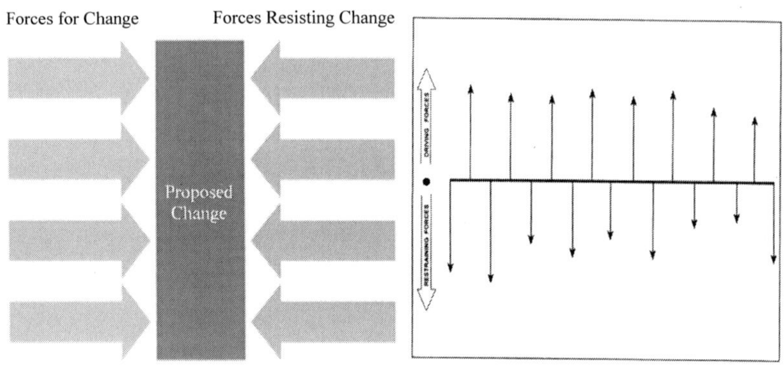

Figure 10.24: Force field analysis.

Utility

Once the positive and negative factors for a change are identified, it will be easier to develop strategy to solve the problems.

xiii) Indigenous Technical Knowledge (ITK)

There are some indigenous knowledge in villages that are used by the rural people to solve their own problems. These knowledge sources are a rich repertoire of culture, practices, beliefs and approaches for understanding and appreciating the rural people. Learning of such ITKs from different sources helps set the context of views expressed by rural people. While collecting the information on ITK, the source, number of people adopting and the results are necessary for their validation.

Discussion Methods

a) Focus group discussions (FGDs)

Focus Group Discussion is centered around a specific problem and enables the villagers for entering into conversation under facilitation of the PRA practitioners.

It is facilitated discussion among a small group of stakeholders designed to obtain perceptions in a defined area of interest. It is the method of rapid assessment and data gathering in which participants congregate to talk about the specific issues and concern based on a list of key themes drawn up by the researcher/facilitator.

For conducting a focus group discussion, a facilitator and an assistant to facilitator are needed. The facilitator leads the group discussion and encourages the participants to offer their perception and views on the key subject without diverting to other subjects. The assistant to the facilitator is to take notes, run the tape recorder, respond to the unexpected interruptions, and is always ready to follow the facilitator's mode of action. Knowledgeable, pleasing personality, politeness, ability to speak local language, respect to local norms and behavior, ethics, patience, etc. are the main criteria of a good facilitator.

A good FGD requires "thinking, listening and managing time at the same time" (Lewis 2003). The moderator should avoid dominating the group and expressing his/her own judgments. Instead, he/she should be open, alert, probe, and encourage everyone to take part in the discussion.

Process

A typical FGD proceeds as follows:

- Start the discussion with an 'ice-breaker,' e.g., a round of introduction of participants;

- Explain the topic for discussion and the procedure to be adopted;
- ensure that all important questions (which may be photocopied and distributed to the participants) are covered, without interrupting the natural flow of the discussion; and use a checklist;
- Record the minutes of discussion and conclusion.

FGDs may one hour to several hours. The participants are asked to structure the discussion in any way they like, which unconsciously projects their own desires, expectations, hopes, fears, and/or repressed wishes into the course of the discussion, if appropriate for the research topic.

b) Semi-Structured Interview (SSI)

Semi-Structured Interviews is a guided interview in which the topics are pre-determined and specific questions are asked during the interview. After the reconnaissance and rapport building activities this technique is undertaken to study village in its entirety. This technique allows village characterization in terms of geographic, demographic, edaphic, farming systems, and socioeconomic features. The PRA team has to conduct a series of 3-4 semi-formal interviews of selected 8–10 key informants (representing village officials, elected leaders, functional leaders, gender, and occupational groups), using a semi-structured interview schedule or checklist.

The key components are:

- Informal and controlled conversation
- Use of a checklist for discussion without using a formal questionnaire
- Use of open-ended questions
- Allows new issues to arise
- Leads to hypothesis

The information and data so collected has to be verified and triangulated with various available secondary sources for authenticity and completeness. This also serves as a baseline, when PRA is conducted as a part of a larger purpose of either village adoption or introduction of development interventions.

Types of SSI

- Group Interviews: The group interview may be focus group (discussed above) or an open group. Outsiders can meet different groups separately or together to know about their problems and opportunities.
- Key Informant Interview: There are many key informants in a village who can share many types of information about rural life. These

informants can be interviewed to gather specific information from them.

- Individual Respondent Interview: Information on household activities, sensitive issues, and representative information can be collected from individual interviews.

Table 10.6: Type of Information collected from different Informants

Type of respondents	Type of information
Model farmers/innovators	Appropriate/advanced technologies
School teacher/educate people	Information relating to education
Officials, researchers	Experience of projects and plans
Village religious leaders	Beliefs, community activities
Women leaders	Women activities, role of women
Health workers	Health, nutrition and sanitation
Old people	History and tradition
Business people	Marketing, processing and storage
Forest officials	Forestry issues

While collecting specific information about a village, the respondents concerned may be interviewed (Table 10.6).

Guidelines

A checklist is used and the participation of farmers is required at equal footing. The facilitator may use probing technologies in case of ambiguous or inappropriate information.

Steps to follow

i) Before the survey:
   - Select the multi-disciplinary survey team.
   - Analyse secondary data.
   - Prepare a checklist for questions in advance.
   - Inform farmers in advance to arrange a meeting at a common place; and
   - Establish note-taking procedures before entering a village.

ii) During a group meeting or individual interview:
   - Start the meeting at a common place giving respect to farmers, their culture and tradition.
   - Do not use the checklist as a questionnaire but use it as a means to stimulate discussion and a participatory dialogue.

- Build the questions to be asked around a list of sub-topics, existing information on the community, or visual material such as diagrams, photographs, or maps.
- Use the "six little helpers" for probing: Who? Why? What? When? Where? How?
- Probe answers carefully and triangulate.
- Take notes during the interview but not excessively and
- Generate new hypotheses.

iii) After the meeting or interview:

- Finish the discussion politely.
- Have evening brainstorming sessions with the team to complete notes and to prepare for the following day.

Common errors in SSI

- Failing to listen closely and repeating the questions.
- Asking lead questions and helping the people in answering.
- Asking vague and insensitive questions.
- Elite bias.
- Failure to probe.
- Being unaware of hypothesis confirmation.
- Prematurely searching for coherence.
- Engaging the informants for a longer period.

A model format of SST is presented in Figure 10.25.

Sequences of PRA

The sequence of PRA depends on the objective of study and kind of data required. Since new methods are being used in PRA, the combination, and sequences of such methods are necessary for in-depth probing. Such sequencing provides an effective rural appraisal. One method leads to another depending on the topic being probed and the objectives.

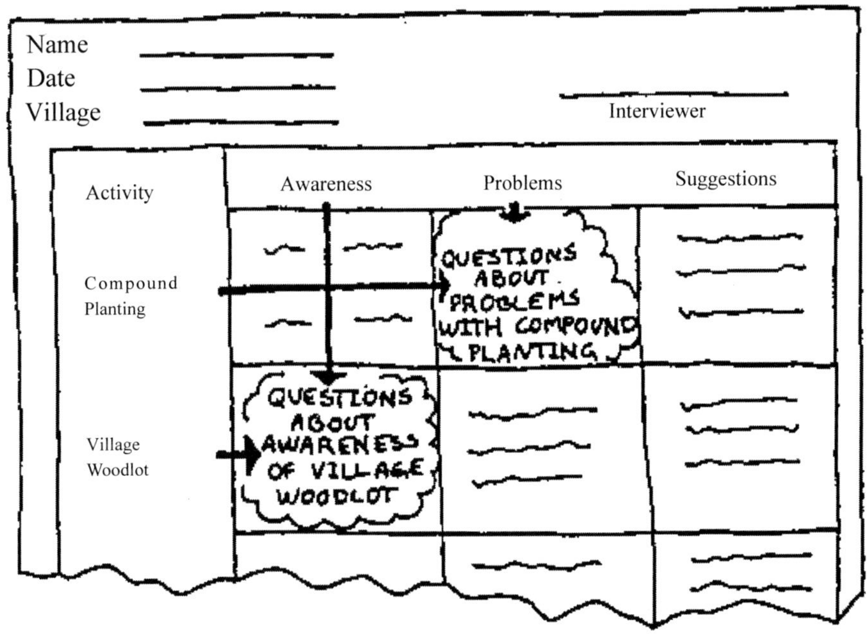

Figure 10.25: Format for SSI.

- With participatory mapping, villagers draw several maps which become successively more detailed and useful. Social map leads to wealth ranking and health map, etc.
- Resource map leads planning transect walks in which the villagers who made the map act as guide for outsiders. The transect leads to discussion and identification of problems and opportunities.
- Social map-wealth ranking-resource map-transect- livelihoods analysis-seasonality calendar are beginning sequences after which other methods can be adopted depending on the objectives, data requirement, and resource availability.
- Sometimes preferential ranking is regarded as a good icebreaker. Individual interviews can follow this.

Although there is no such hard and first rule on sequencing of PRA exercises, it is desirable to follow a sequence to make the exercise cost-effective and generating greater information in the process. For example, a social map, wealth ranking, and resource map can follow a geographical transect. This can be followed by livelihoods analysis, seasonality diagram, technology map, treatment map, scoring and ranking, etc. to solve the problems of villagers. Community problem analysis, focus group discussion, SWOT analysis, time line, time trend, Venn diagram, cause and effect diagram, etc. would be required for planning purpose. For forest planning and natural resource management, a resource map and geographical transect can be followed by food calendar, and matrix ranking.

As general rules of thumb, mapping, and modeling are good techniques to start with because they involve several people, stimulate, much discussion and enthusiasm among people. Maps and modeling can be followed by transects involving some people who have constructed the maps. These are followed by group methods.

Participatory Learning and Action (PLA)

What is Participatory Learning and Action?

Participatory Learning and Action (PLA) is a community-based participatory methodology used for qualitative research. Through PLA local people can explore and share their knowledge, and plan and carry out actions to effect change within their communities. It is a creative learning process by local people with the skills and confidence to plan their work. It breaks down barriers between community representatives and professionals and promotes a shared understanding of each other's priorities and constraints. The main purpose of PLA is to facilitate rural people to analyse their own situation, rather than have it analyzed by outsiders, and to ensure that any learning is then translated into action (Gosling and Edwards 2003). PRA was changed to Participatory Learning and Action (PLA) to reflect its broader application.

Philosophy of PLA

PLA is a way of thinking that emphasizes reversals in power relations between communities and outsiders (such as researchers, evaluators or program planners).

PLA uses almost all the tools and techniques used in PRA. In PLA the outsiders learn from the insiders of village. The communities are supported by outside facilitators to analyse their own situation, make decisions about how to best tackle their problems, and, as a result, feel empowered to take action.

Principles of PLA

In principle PLA is highly participative and inter active, inclusive, a process of learning, community led, action oriented, flexible and adaptable. PLA is based around group analysis and learning for action.

- It is flexible, adaptive, and innovative, rather than adopting top-down or rigid methods of data collection and analysis.

- It encourages people to discuss issues, invites errors and mistakes in a non-judgmental environment.
- The language and concepts of PLA should be appropriate to the local culture and context.
- Seeks out voices and opinions of the most marginalized communities and people, such as children, women, people of lower caste or status, or people with disabilities.
- Ensures triangulation by collecting information about in different ways, and from different sources to make it reliable and accept different viewpoints.
- Requires multi-disciplinary teams to facilitate, including people from all communities with different skills and different views.
- Provides insights and understanding that helps guide community development, rather than providing academic data.

Ground Rules

- Give everyone the opportunity to participate
- Respect what people say
- Don't dominate
- Be on the same level as the participants
- Don't make assumptions
- Don't rush
- Hand over the pen'
- Learn from mistakes
- Be flexible
- Make it Fun

Process

PLA follows four distinct phases:

- Training, research and data collection
- Analysis of action/community planning
- Verification of report and draft plan
- Collective action

How it Works?

The goals and objectives of the PLA work are clarified first and agreed with the communities. If there are multiple or conflicting goals then issues are resolved before proceeding. A set of relevant tools and approaches are selected after that.

PLA comprises many different tools and approaches, some of which are described in the next section. The tools and approaches should be applied in a participatory manner. In addition to the exercise being carried out by the trained facilitator, the information requires triangulation before validation by the community.

Two important concepts in PLA are optimal ignorance and appropriate imprecision (Chambers 1983). The people may be asked to collect and analyze just the required data.

When the point of optimal ignorance is achieved, the next step is to develop a community action plan.

PLA Tools and Techniques

The methods and techniques used in PLA include direct or participant observation, semi-structured interviews, focus group discussions, photography, video, case studies, and the use of secondary data sources. Most of the tools and techniques of PLA are similar to that of PRA. However, the following tools and techniques are used in PLA.

PLA Tools

A. Early stages B. Analysis C. Prioritization and Action Planning

A. Early stages

- Mapping (Transect map, Social Map, Mobility Map)
- Timeline
- Time Trend

Seasonality Calendar

B. Analysis

- Spider diagram
- Venn Diagram
- Flow Diagram
- Force field analysis
- Causal Impact Diagram

C. Prioritization

- Ranking and scoring (wealth ranking, pair-wise ranking, matrix ranking)
- Semi-Structured Interview

PLA requires good facilitation skills for which the role is very important. The facilitator needs to make sure that PLA is carried out to high quality and ethical standards as the attitudes and behaviors are conducive to learning.

References

Chambers, R. 1992. Rural appraisal: rapid, relaxed and Participatory, IDS Discussion paper No. 331. Institute of Development Studies, U.K.

- (1994). The Origins and Practice of Participatory Rural Appraisal. World Development. 22(7): 953-69.

- (1997) Whose Reality Counts. Intermediate Technology Publications, London.

- (2004). Notes for Participants in PRA-PLA Familiarisation workshops in 2004. Institute of Development Studies, U.K.

Frcire, Paulo, 1968. Pedagogy of the Oppressed (New York: The , Seabury Press).

Gosling, L. and Edwards, M. 2003. Toolkits: A Practical Guide to Assessment, Monitoring, Review and Evaluation., 2nd edition. UK: Save the Children, UK.

Lewin, K. 1948. Resolving social conflicts; selected papers on group dynamics. Gertrude W. Lewin (ed.). New York: Harper & Row.

Peters, Tom, 1989. Thriving on Chaos: Handbook for a Management Revolution (London: Pan Books.

Ritchie, J. and Lewis, J. 2003. Qualitative Research Practice—A Guide for Social Science Students and Researchers. London, Thousand Oaks, CA: Sage Publications Ltd.

Robert Chambers, R. 1992. Rural Appraisal (Rapid, Relaxed and Participatory); Sussex: Institute of Development Studies, Sussex.

Suggested Reading

Chambers, R 1983. Rural Development: Putting the last first. Longman inc., USA, 1983.

Government of India, 2011. "Common Guidelines for Watershed Development Projects – 2008" (Revised edition 2011). National Rainfed Area Authority, Planning Commission. New Delhi, India.

Mukherji, N. (1993): Participatory Rural Appraisal; Methodology and Applications.

Mukherjee. Ncela, 1992. "Villagers' perceptions of rural poverty through the mapping methods of PRA," RRA Notes, No. (15) 21-6.

National Rainfed Area Authority. 2011. "Common guidelines for watershed development projects-2008", Revised editions-2011, National rainfed area authority, Planning commission, 2011.

Peasant Association Panning for Natural Resource Management, IIE London.

Pretty JN and Simplice DV. 1997. Using rapid or participatory rural appraisal. In: Improving agricultural extension: A reference manual. FAO (Food and Agriculture Organization of the United Nations), Rome, Italy.

Pretty JN, Guijt I, Thomson J and Scoones. 1995. Participatory learning and action; a trainers guide. IIED (International Institute for Environment and Development), London, UK.

Scoones, I. and J. McCracken, 1989. Participatory Rapid Rural Apaisal in Wollo.

Sontaki Bharat, S. et al. 2019. Participatory Rural Appraisal (PRA) Tools and Techniques.

11

Decentralized Planning and Micro Level Planning

Introduction

Decentralization means the action or fact of reducing or undoing "centralization" or "concentration." The dictionary meaning of decentralization is "The process of transferring the functions from the Central Government to the local units." Decentralization is devolving power of decision–making with the lower level of organization. In a decentralization, delegation and devolution of power are given to the lower level. Devolution implies a situation in which lower-ranking units acquire greater autonomy in respect of certain defined functions, including decision-making authority. Devolution of functions, powers, and responsibilities are normally granted through legislation.

Decentralized Planning in India

With the introduction of the Community Development Programme (CDP) and National Extension Service (NES) in 1952, increasing agricultural production had been on India's agenda in different plan periods. The success of CDP and NES was reviewed by the Balawant Rai Committee in 1957 and the committee realized that community development can be achieved only when the community understands its problems, owns its responsibilities, exercises the necessary powers through its elected representatives, and maintains a constant and intelligent vigilance on local administration. The recommendation of the committee led to the preparation of plans at the level of state, district, block, towns, and Panchayats in the second plan period from 1957 to 1962. The Fourth Evaluation Report (Programme Evaluation Organization, Planning Commission) in 1959 recommended that organized and institutional participation by the public in development programs is bound to give better, more consistent, and more lasting results. In 1977, Ashok Meheta Committee suggested changing the concept of Panchayati Raj to a political institution. In 1981, the eminent economist Prof. M.L. Dantewala observed that the merit of decentralized planning is derived from its ability to discover dormant resources and skills and its endeavor to activate them. However, there was slow progress in decentralized planning up to the eighth plan (1992-97). The chronology of committees reviewed the progress and suggested for decentralized planning is given in Table 11.1.

Table 11.1: Committees Suggested Decentralization in India

Year	Item	Ideas and Concepts
First Plan 1951-56	Community Development Blocks	To break up the planning exercise into national, state, district, block, and local community levels
Second Plan 1956-61	District Development Councils	Drawing up village plans and popular participation in planning through the process of democratic decentralization
1957	Balwant Rai Mehta Committee	Village, Block, and District Panchayat institutions established
1967	Administrative Reforms Commission	Resources to be given/local variations accommodated, purposeful plan for the area
1969	Planning Commission	Formulated Guidelines; detailed the concept of the district plan and methodology of drawing up such a plan in the framework of annual plans, medium–sized plans, and perspective plans
1978	Prof. M.L. Dantewala	Intermediate Panchayat level planning to form a link between village and district-level planning
1983-84	CSS/Reserve Bank of India	Strengthen district plan/district credit plan
1984	Hanumantha Rai Committee	Decentralization of functions, powers, and finances; setting up of district planning bodies and district planning cells
1985	GVK Rao Committee	Administrative arrangements for rural development; district Panchayat to manage all development programs

A historic development was made by the introduction of the 73rd and 74th Constitutional Amendment Act, 1992 for the establishment of local self-governance in rural and urban areas respectively. Emphasis was given on creating a representative local institution for "area planning for economic development and social justice." The 73rd Amendment has mandated the creation of a District Planning Committee (DPC) to undertake planning at the district level. DPC is required to integrate the micro-level area plans emanating from village Panchayats and block Panchayats. A committee set up by the National Development Council in 1993 recommended micro-planning and involvement of people at a grassroots level as an effective means through which economic development can meet the aspirations of the people and ensure that benefits really percolate down to the lowest strata. Ideally, the micro-level planning unit for meeting the needs of the individuals and the families has to be at the village level. Planning for economic development and social justice is a mandated function of local governments and the District Planning Committee (DPC) has a crucial role to play in the process. Under Article 243ZD of the Constitution of India, a District Planning Committee shall be constituted at the district level in every State to consolidate the plans prepared by the Panchayats and the Municipalities in the district and to prepare a draft development plan for the district. In preparing the draft development plan, the DPC shall have regard to matters of common interest between the Panchayats and Urban Local Bodies (ULBs) for spatial planning, sharing of natural resources, the integrated development of infrastructure and environmental conservation, and the extent and type of available resources, both financial and otherwise. The DPC in this endeavor is also mandated to consult such institutions and organizations as may be specified.

The Article-243(G) of the 73rd Constitutional Amendment Act-1992 mandates the nodal role of Gram Panchayat with regard to local area planning which is an indispensable pre-condition for GP to function as an "institution of local governance" and policy formulation at the grass root, not as a mere implementing agency for plans and programs. The 73rd constitutional amendment enjoins that the village Panchayat will have to perform basically two functions: a) the preparation of plans for economic development and social justice; b) the implementation of schemes for economic development and social justice as may be entrusted to them including those in relation to the matters listed in the 11th Schedule of CCA, 1992. The twenty-nine matters as per the 11th Schedule are indicated in Table 11.2.

Participative Rural Planning

Panchayat has to identify the felt needs and priorities of local people, map its own resource envelope, prepare need-based projects, and allocate resources judiciously based on participatory planning at the local level. The Gram Panchayat also needs to identify areas for which a long-term perspective like infrastructure, market development, provision of basic services such as drinking water, sanitation & street lighting at the village level, plantation, livelihood support to poor sections and welfare of vulnerable groups are to be included in the Gram Panchayat plans.

Table 11.2: Subjects of Panchayat as per Eleventh Schedule (Article 243 G)

S. No	Subjects
1	Agriculture, including agricultural extension
2	Land improvement, implementation of land reforms, land consolidation, and soil conservation
3	Minor irrigation, water management, and watershed development
4	Animal husbandry, dairying, and poultry
5	Fisheries
6	Social forestry and farm forestry
7	Minor forest produce
8	Small-scale industries, including food processing industries
9	Khadi, village, and cottage industries
10	Rural housing
11	Drinking water
12	Fuel and fodder
13	Roads, culverts, bridges, ferries, waterways, and other means of communication
14	Rural electrification, including the distribution of electricity
15	Non-conventional energy sources
16	Poverty alleviation program
17	Education, including primary and secondary schools
18	Technical training and vocational education.
19	Adult and non-formal education
20	Libraries
21	Cultural activities
22	Markets and fairs
23	Health and sanitation, including hospitals, primary health centers, and dispensaries
24	Family welfare
25	Women and child development
26	Social welfare, including the welfare of the handicapped and mentally retarded
27	Welfare of the weaker sections, and in particular, of the Scheduled Castes and the Scheduled Tribes
28	Public distribution system
29	Maintenance of community assets

The stock-taking exercise on infrastructure, civic amenities, economic development, social development, and natural resources has to be done to identify the gaps. Based on the stock-taking information and analysis of gaps and deficiencies, the sector-wise long-term priorities are to be worked out perceptively for a period of five years by the GP planning unit. From the perspective of long-term priorities, immediate priorities for the current year are to be identified. The felt needs are required to be prioritized by using prioritization matrix. The prioritization made by the GP level planning unit is to be examined by the Gram Panchayat body. After necessary consideration by the Gram Panchayat, all the proposals are to be placed before the Gram Sabha for final approval. The GP plan is then sent to the intermediate Panchayat (Block Panchayat) for integration.

Planning Methodology

Planning at the GP, block, and district level is a continuous process that is based on field situation, empirical analysis, program design, and management of implementation. Participatory planning is formulated as per the felt needs of people and to tap the local knowledge in the planning process. The outcomes and achievements of one plan become the input for the next plan. The planning cycle is divided into five stages:

- Situation analysis.
- GP/Block/District diagnosis.
- Program design.
- Implementation.
- Monitoring and evaluation.

The planning exercise starts from the district stocktaking exercise to assess the level of human development and the availability of natural resources, financial resources, and infrastructure. The envisioning and perspective planning process follows the stocktaking exercise and a five-year perspective plan and a 15-year vision is worked out. Financial resource mapping is done at the district level and the budget and fund envelope is communicated to all the planning units. After getting the stock-taking document, budget envelope, and 15-year vision participative rural planning and participatory urban planning within the district are done.

Situation Analysis

Situation analysis of Gram Panchayat (or block/district as the case may be) is required to understand the ground realities. Data based on social, cultural, and economic spheres is essential for planning and implementation. Situation analysis involves data collection, data analysis, identification of potentials, problems, needs, and resource mapping

The data can be classified as village profile and household survey. The village profile covers natural resources, human resources, infrastructure, services, industry, etc. A Household survey is done to know the income, assets, educational status disparities among groups, etc. The type of information to be collected in the village profile and household survey are:

Village profile: Demographic data (population, employment BPL families), agriculture (cropping pattern, gross cropped area, production, technology adopted, problems), natural resource management (land holdings, land use pattern, common property resources, problematic areas, forest), irrigation (sources, area under irrigation, potential, modes of irrigation), animal husbandry (population, production, problems), fisheries (number of water bodies, production, a system of fishery, problems), forest (area under reserved forest, village forest, agroforestry, joint forest management, minor forest produces), rural connectivity, housing, marketing, energy, industries, education infrastructure, health and sanitation, drinking water, women and child development, social welfare, civic amenities, poverty alleviation and employment, and public distribution system.

Table 11.3: Resource Matrix of Panchayat

S.No	Type	Item	Unit km/number	Present status
1	Economic infrastructure	Road Irrigation Rural market Godown Bank Post Office Transport		
2	Social infrastructure	Hospitals CHC PHC Sub-center Schools Primary Middle High school College Drinking water Community center AWC Training infrastructure		
3	Natural resources	Land resource Forest area Water resources		

(NB: The list is indicative and not exhaustive)

Data analysis: The collection of data follows analysis to know the present status, comparative position with other regions, and past years. It is also necessary to identify the inter-group and intra-group disparities. Comparative analysis, analysis of disparities, and trend analysis are usually done. Certain indicators like population growth, poverty, employment, agricultural production, literacy rate, etc. help understand the direction of present development programs.

Identification of potentials, problems, and needs: A resource matrix is prepared after data analysis in order to facilitate possible action to be taken for future planning. Resource matrix covers economic infrastructure (road, irrigation, rural market, go-downs, banks, post office, and transport), social infrastructures (hospitals, schools) drinking water, community center, Anganwadi center, training infrastructure) and natural resources (forest area, land, and water resources). An indicative resource matrix is presented in Table 11.3.

Potential areas include irrigation, road infrastructure, human resource development, income, forest, industry, etc. An indicative list of potential areas has been given in Table 11.4.

Problems and constraints are to be identified and prioritized. At the same time, the target groups such as smallholders, tribals, women and children, and rural poor are to be identified for development programs. An indicative problem analysis statement is presented in Table 11.5.

Resource mapping can be done by using PRA techniques, aerial photo interpreting, and Geographic Information System (GIS).

Methodology for Village Development Plan

The village development plan should take into consideration all the areas falling under the revenue boundary of the village including areas of the main village or hamlets, farmlands, and areas under other uses. Since the people are to be involved in the village development plan, the methodology should be simple so that people can easily understand and participate. The following methodology is suggested for the village development plan.

Table 11.4: Identification of Potential Areas in Panchayat

S.No	Item	Resources	Potential	Future scope
1	Irrigation	Canal MIP	Area can be irrigated	Increasing crop production
2	Road infrastructure	Metallic All weather	Marketing network	Improving marketing network
3	HRD	Availability of skilled manpower	Required for industry	Entrepreneur development

4	Income	High-income people	New technology for agriculture and industries	Boosting agro-based industry and higher growth
5	Forest	Forest products like bamboo and timber	For forest-based industries	Productive employment

(NB: The list is indicative and not exhaustive)

Table 11.5: Example of Problem Analysis in Panchayat

Category	Sector	Type	Problem	Causes/ constraints
Social	Education	Individual household Infrastructure	Low education Lack of building	Lack of awareness Lack of funds
Economic	Poverty alleviation Rural connectivity	Household Infrastructure	Low income Bad roads	Lack of employment Lack of funds
Service	Marketing	Infrastructure	Lack of godown Lack of transport	Lack of funds Lack of funds
Ecology	Natural resource management	Land management Forest Management Water management	Soil erosion Denudation Want of proper utilization	Unscientific cultivation Felling of trees Unscientific water use

(NB: The list is indicative and not exhaustive)

Formulation of objective: The primary objective is to reflect people's needs and priorities. The common areas of development should be poverty alleviation and employment generation, community welfare, infrastructure development, economic development, social development, human development, and natural resource management.

Data collection: The basic data should cover demography, environmental support system, social and physical infrastructure, maps, plans, and details of development programs carried out or currently going on in the village.

Preparation of inventory and thematic maps: The preparation of an inventory of the resources and spatial maps is to be prepared with the latest cartographic techniques by the professionals. The information should be understood by the villagers and they are used effectively in the analysis to be undertaken in the next step.

Study of development status, dynamics of change, projections, potentials, and issues: Data analysis needs to be undertaken by professionals and explained to the villagers to make them understand the problems. The people should understand the short-term and long-term needs, problems and potentials, and issues and thrust areas for development.

Preparing Village Development Plan: The village development plan should include (i) a 15-20 years vision, (ii) a perspective plan for five years, and (iii) the annual action plan drawn from the perspective plan. The annual plan should be reviewed every year for better performance next year. These annual reviews should be helpful in drawing up a new five-year plan. These three plans are interlinked in space and time with higher level plans at the block and district levels. The difference between a five-year perspective plan and 15 years vision are shown in Table 11.6.

Table 11.6: Five Year Plan vs 15 Years Vision

Five-year Perspective Plan	**Fifteen-years Vision**
As per budget allocation for five years	Indicates a larger picture and sets out broad development goals without finding constraints
Guided by existing funding priorities and schemes	Aims at influencing funding priorities over a long period
Projectized and implemented through annual plans for prioritized activities	Sets the priorities without being in projectable mode

Usually, the District Planning Committee prepares the district vision while perspective plans are prepared at each level of Panchayat.

Micro-Level Planning

Micro-level planning is intended to bring about integrated area development in the countryside. It is a bottom-up planning in which the needs of the local areas as well as the purpose of regional development are served by such planning. The location of specific socio-economic activities and their interlinkages over a region are major concerns of micro-planning. The national plans provide a broad framework for strategic development while a micro-level plan focuses on the proprictics of different regions depending on their specific need. However, both the 'macro' and 'micro' level planning are complimentary to each other and their combined use is required for the overall socio-economic development of the country. To make any development planning effective, there is a need to follow a two-pronged approach working simultaneously and in a coordinated way from the grassroots level up, and the national level down (Singh, 1999: 247).

The concept of micro-planning emerged taking into view the emerging socio-economic conditions of the country and the inadequacy of past planning efforts in checking regional disparities. Micro Planning is a community-based empowering process for preparing a road map for the development and management of infrastructure and livelihood enhancement of the communities with properly defined roles and responsibilities. It is required for the integrated and balanced development of an area. In a nutshell, micro-level planning is concerned with the allocation of resources of the planning entity concerned, to maximize whatever goals the entity may have (Singh 1999: 246).

The Fourth Plan envisaged the necessity of micro-planning in rejuvenating an underdeveloped area by proper utilization of the natural and human resources. It stressed upon micro-planning at district and lower levels with an area development program to provide infrastructure and other growth requirements in each area. In micro-level planning, four prime considerations are (i) identification of the specific needs of the area, (ii) an accurate assessment of the limits and opportunities imposed on available resources of the area, (iii) selection of suitable strategy for development, and (iv) proper coordination at multiple spatial levels.

The Concept, Need, and Objectives of Micro-planning

The Concept

The term micro-planning or area planning implies a multi-level and decentralized planning approach to the overall development of a region. It is essentially a spatial development planning that tends to utilize all kinds of available resources – natural, human, and others to the fullest extent to minimize the socio-economic imbalances.

The Need

A micro-planning, in its true perspective, tends to be much more responsive to the emerging socio-economic problems at various territorial levels. It is suggested for the all-round socio-economic development of a geographically diverse country like India against the single national level sectoral planning. It helps overcome the possibility of regional disparity in fostering economic growth and development. In order to carry the benefits of development to the poor, to ensure the continuity of balanced growth, and to provide social justice, micro-level planning has been recommended.

Aim and Objective

The aim of 'micro-planning' is planning from the lowest level i.e., from the functional community upward to a clearly defined region to fulfill the needs of the local areas and ensure the process of integration of the different areas with an objective to attain balanced regional development. Therefore, the location of specific socio-economic activities and their interlinkage over a region or particular geographical area are the major objectives of micro-level planning. Micro-planning lays greater emphasis on those sectors which support the poor and weaker sections of the community in rural areas with an aim to offer them a better quality of life and improve their socio-economic conditions. Therefore, emphasis is given to spatial types of economic, social, and environmental plans to alleviate rural poverty and inequality through micro-level planning.

The Background of Micro-Planning in India

During the first two plans, there was no effort for micro-planning. In the Third Plan, the regional focus in planning became more explicit, and for the first time, it gave serious thought to the problems of regional development. The Fourth Five-Year Plan (1969-1974) emphasized micro-planning to correct some of these regional imbalances on the assumption that plans made at the national and state levels can be brought down to the people of lower levels in a much more efficient manner. The Fourth Plan put considerable importance on planning at the district level and on experimental studies on 'growth centers' for evolving an appropriate micro-planning strategy at the grass-root level (Fourth Five Year Plan 1969: 229-30). In fact, the Fourth Plan initiated micro/regional planning from the grass-roots under the name of area development taking the regional resource potentialities and limitations into consideration.

Approaches to Sustainable Development

Recognizing the fact that macro-level planning cannot address local variations in resources and needs, an area development framework drawn up at the district and block levels was considered to be more realistic than one formulated at the state level. Therefore, for micro-level regional planning, the district was selected as a planning unit. With the aim to accelerate the development of backward areas and to reduce regional disparities in socio-economic advancement, the concept of integrated area development had emerged for sustained development of the targeted 'area'. Integrated area development refers to the appropriate location of social and economic activities over a physical space for the balanced development of a particular region.

Approach and Strategies for Micro-planning

As stated above, micro-planning is planning from the lowest level upward to a clearly defined area or region. In many cases, this region may be co-terminus with the district. In this planning, the needs of the local areas as well as the purpose of regional development are served. The location of specific socio-economic activities and their interlinkages over a region are major concerns of micro-planning. However, for all practical purposes, both the 'macro' and 'micro' methods of planning are complementary to each other, and their combined use is essential in bringing about the overall socio-economic development of the country. It is necessary to follow a two-pronged approach for the coordinated way of grassroots planning and national-level planning.

Micro-planning offers planning within a spatial framework so that all kinds of resources and endowments may be utilized fully and the results could be equally shared socially. The approach of micro-planning led to area-specific plans as in the case of Command Area Development (CAD), Desert Development Programme (DDP), Drought-Prone Area Development Programme (DPAP), Integrated Area Development Programme (IADP), Hill Area Development Programme (HADP), Tribal Area Development Programme (TADP) and Whole Village Development Programme (WVDP). They are directed at specific areas or regions which suffer from some disabilities or are prone to certain hardships.

Elements of Micro-planning

The elements of micro-planning are rapport building, situation analysis, vision and goal setting, strategic framework, implementation, monitoring, and evaluation.

Rapport building: It is a close and harmonious relationship in which the people understand each other's feelings or ideas, and communicate smoothly. The special team enters into rapport with communities by having healthy interactions.

Situation analysis & data collection: It is an in-depth analysis of the family situation, individual-centric data collection regarding current details, major issues, weaknesses, and challenges. Usually, information on geo-political situation, socio-demographic situation, demographic profile, composition of social group, public infrastructure and services, public health information, and socio-economic information are collected for situation analysis. These data are also required in the envisioning exercise.

Vision & goal: They are a critical element of a micro plan. Vision defines what the achievements will be in the future. The vision reflects the community's history, its core values, and competitive strengths. Goals are tangible milestones to be achieved to realize the vision. The targets in the micro plan are measurable, specific, relevant, achievable, and time-bound.

Strategic framework: It comprises the strategic mechanisms to accomplish the goals and objectives set by the authority. The micro plan specifies and prioritizes initiatives and aligns resources.

Implementation: It is the performance of the micro plan based on effective execution. Systems, principles, and accountability measures are put in place for carrying out the micro plan.

Monitoring: Monitoring is required to assess the progress of the plan against agreed targets, and goals to identify necessary changes, assess options and constraints, and to re-define realistic objectives.

Evaluation: These are systematic assessments of programs or policies, their design, implementation, and results, with the aim of determining their relevance, efficiency, effectiveness, impact, and sustainability with regard to overall government priorities. The monitoring and evaluation of plans have been shown in Figure 11.1.

Methodology for Micro-planning

It begins with the district-level stakeholders meeting, team mobilization, survey, data collection, preparation of plan, presentation, and approval after obtaining the feedback (Figure 11.2).

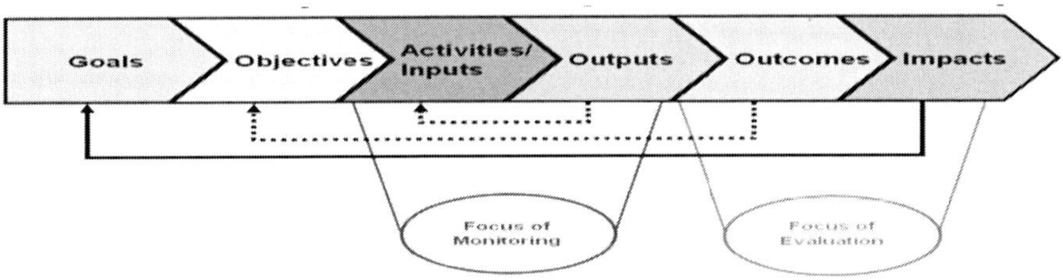

Figure 11.1: Monitoring & evaluation of plans.

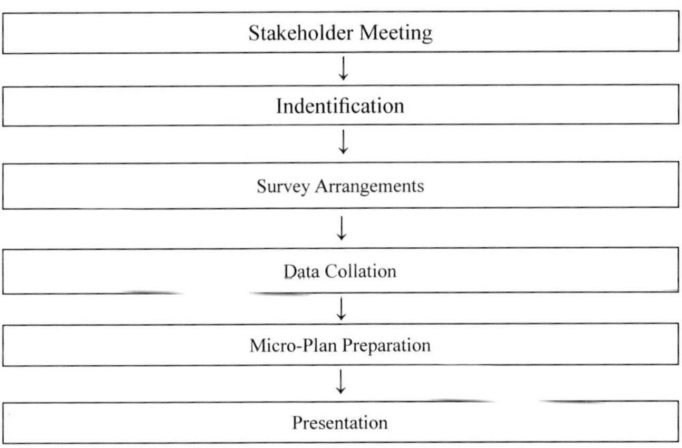

Figure 11.2: Methodology of micro-planning in a district.

The following steps are taken for the preparation of a village micro-plan.

- A five-day long exercise in a particular village is done by a team of village planning facilitators (NGOs/CBOs or Planning Team of GP) depending upon the population size and spread of the village.

- The team interacts with village leaders and volunteers during the preparation of the micro-plan of the villages.
- The village micro-plan is prepared in consultation with the village leaders, and standing committee members of GP and presented to the members
- Block level/GP level stakeholders meeting and discussion regarding the micro plan preparation. Cross-checking of data based on a group discussion with the line department at the field level. Data cleaning and data verification are also possible in the stakeholders' meetings.
- Identification of families including vulnerable families and beneficiaries of government schemes in the village.
- Scheduling of methodology for a family survey and arranging of survey formats for baseline data collection.
- Participatory collection through the use of PRA/PLA tools for situation analysis, need assessment, problem identification/ prioritization, resource analysis, etc. for micro plan preparation.
- Preparation of a micro-plan of the village in consultation with standing committee members
- Convening Gram Sabha by the GP and presentation of micro-plan to Gram Sabha members for approval.

Five-day Schedule

- Day one: Stakeholder meeting, village rapport building
- Day two: Stock-taking exercise of the village and database creation
- Day three: Participatory data collection, analysis of problems/needs/ resources
- Day four: SWOT analysis, prioritization of needs/problems, and finding alternate solutions to problems
- Day five: Preparation of Micro-plan in consultation with village leaders, standing committee, and community.

Baseline Data Collection

- Use a survey questionnaire for the HH Survey.
- Collect data on demography, natural resources, infrastructure availability, livelihoods, important schemes in operation, economic activities, agriculture details, health, drinking water, sanitation, education, basic amenities, institutions, SHGs, etc. for stock-taking.

- Demography: The information on various aspects such as population, households, occupational profile, economic category of farmers, land use &crop land pattern, irrigation status, etc.
- Economic Status of Villagers: The information on assets such as the status of houses (Thatched roof/ Asbestos roof/ Pucca/RCC etc.), electrification, possession of livestock and mechanized farm implements, current income generating activities such as crop production, Livestock (milk, goat, sheep, poultry), Fishery, Small trading, Shop keeping, traditional artisan work, etc.
- Village Resources/ Infrastructure: Detailed information on various types of Infrastructure/amenities such as schools, college, hospitals, post offices, etc., and their location, distance from the village and condition & status of present use

PRA Steps

- Situation analysis.
- Tool to tool cross-checking.
- List out problems.
- Prioritize.
- Problem analysis.
- Options/solutions.
- Option analysis – based on productivity, stability, and equity.
- Option assessment.

PRA/PLA Tools to be Used

- Participatory Mapping (social, resource, and technology).
- Wealth Ranking.
- Time trend and Timeline.
- Transect Walk.
- Venn Diagram.
- Raking and Scoring.
- Seasonality Calendar.
- Problem Cause Tree.
- Impact Diagram.
- Community Problem Analysis.

Situation Analysis and Livelihood Analysis

- Livelihood baseline.
- Livelihoods impact appraisal.
- Livelihood assessment.
- Sustainable livelihood framework (Figure 11.3).

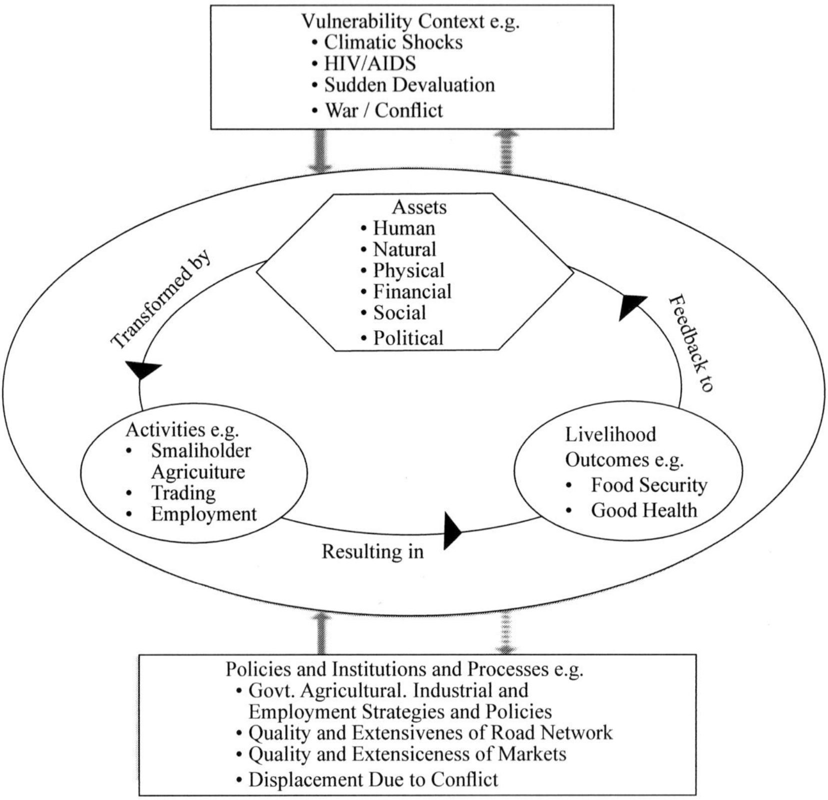

Figure 11.3: Sustainable livelihoods framework.

SWOT Analysis

A SWOT analysis is designed to facilitate a realistic, fact-based, data-driven look at the strengths and weaknesses along with opportunities and threats. SWOT analysis is very important for the district envisioning process.

Elements of SWOT

- Strength: Resource or capacity that can be used effectively to achieve its objectives.

- Opportunity: Any favorable situation in the environment.
- Weakness: Limitation, fault, or defect that will keep it from achieving its objectives.
- Threat: Unfavorable situation in the environment that is potentially hampering the success (Figure 11.4).

Actions required: BUILD on strengths, ELIMINATE weaknesses, EXPLOIT opportunities, and MITIGATE the effect of threat.

Some Techniques of Good Micro-Planning

As per the Manual of Integrated District Planning, 2008 (Planning Commission, Government of India), a five-day exercise is required for preparing a village micro-plan (Table 11.7).

Table 11.7: A Five-Days Schedule for Preparing Micro-plan

Day	Activities
1	Environmental building and village overview: Meeting with village leaders, and local institutional heads, and meet the Gram Sabha.
2	Creation of village database: participatory mapping, household survey, and assessment of infrastructures and services
3	Data analysis and identification of issues; use of appropriate PRA tools and techniques and triangulation with secondary data.
4	Visualizing solutions and prioritizing strategies: FGD, review of works done, and discussion with Gram Sabha
5	Formulation of Village Plan and Sharing: Resource analysis, projectization, budgeting, and the implementing plan. Presentation of plan in Gram Sabha for approval.

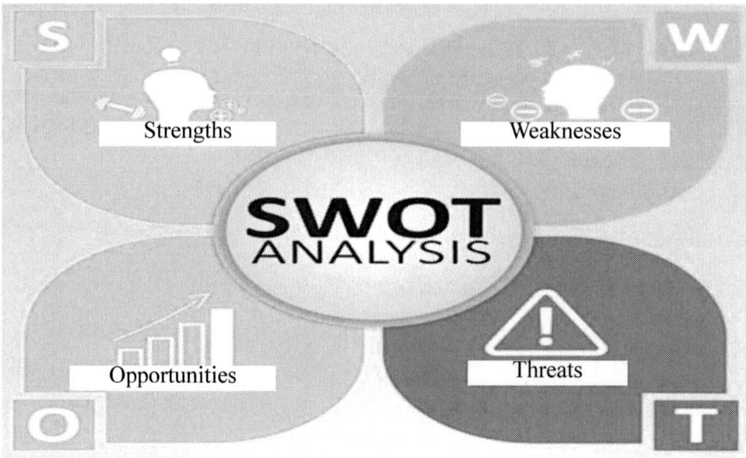

Figure 11.4: Elements of SWOT

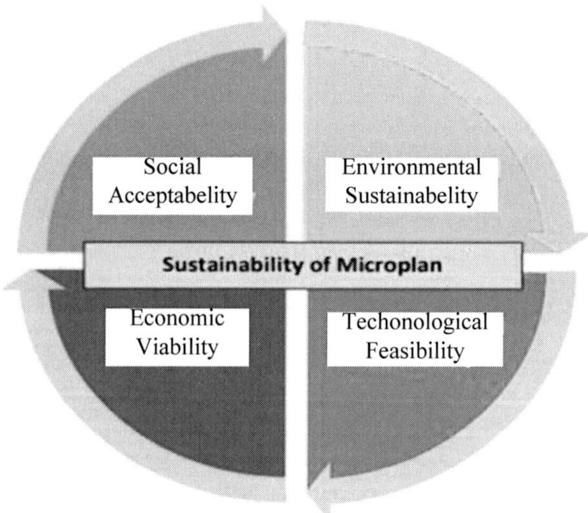

Figure 11.5: Sustainability of micro-level plans

Benefits of MLP

- Problem-solving and decision-making on the basis of alternatives
- Helps find the root cause of problems
- Better understanding of the problems of the communities, analysis, and finding solutions

- Identification of resources in the village and planning for optimum utilization
- Identification of needs and their prioritization
- Ensuring participation of the communities in preparing their own plan and inclusion of vulnerable groups in the village
- Community empowerment

Sustainability of MLP

The MLP should have social acceptability, economic viability, environmental sustainability, and technological feasibility (Figure 11.5).

References

Government of India, (1992a): The Constitution (73rd) Amendment Act, GOI, New Delhi.

Ministry of Panchayati Raj (2006.), Planning at the Grassroots Level: An Action Programme for the Eleventh Five Year Plan, Government of India, New Delhi.

Planning Commission of India (2008): Manual for Integrated District planning.

Ramachandran, V (2006): Expert Group Report on Grassroots Planning.

Singh, K. (1999): Rural Development: Principles, Policies and Management, PP-246-247.

Suggested Reading

Gadgil, D.R. 1967. District Development Planning. Gokhale Institute of Politics and Economics.

Government of India, Planning Commission 1984. Report of the Working Group on District Planning.

Govt. of India, Planning Commission 1957. Report of the Team for the Study of community Projects and National Extension Service, Committee on Plan Projects, New Delhi, Vol. I, 1957.

Gupta R.C. 1992. Integrated District Development Plan for Solan District Institute of Public Administration, Himachal Pradesh (mimeo).

Jain, L.C 1996. Decentralised Planning for Agriculture and Rural Development; NABARD.

K.V.Sundaram, 1997. Decentralised Multi-level Planning Principles & Practice: Concept Publishing House, New Delhi.

Lakadawala, D.T., "`Panchayati Raj in Karnataka today: 1994. Its National Dimensions, Quoted in Decentralized Planning and Panchayati Raj (Proceedings of the D.T. Lakdawala Memorial Symposium).", New Delhi: Institute of Social Sciences, New Delhi, and New Delhi: Concept Publishing Company, New Delhi.

Mishra, R.P and Achyutha, R.N. 1989. Micro-level Rural Planning.

Patna Singh, Tarlok 1969. Towards on Integrated society: Reflections on Planning, Social Policy and Rural Institutions. Orient Longman: Bombay.

Planning Commission 1984. Report of the Working Group on District Planning Vol. 1 & 2, GOI, New Delhi.

Ramachandran, V. 1988. Report on the measures to be taken for democratic decentralization at the district and lower levels', Vol. I, Government of Kerala, Trivandrun.

Sahay B.N. 1995. The Planning Process - A Synoptic Review Paper from National Workshop on Implication of Constitutional Amendments on District Planning at S.PA., New Delhi.

12

Gender Mainstreaming for Rural Development

Gender Concepts

Gender is different from sex. **Sex is the** *biological and physiological characteristics that define humans as female or male. Gender refers to* **the** *social attributes and opportunities associated with being female and male and to the relationships between women and men and girls and boys.* Sex symbolizes biological differences between men and women whereas gender is recognized as a social characteristic that cuts across caste, class, occupation, age, and ethnicity. Gender identifies the social relations between men and women and gender differentiates the roles, responsibilities, resources, constraints, and opportunities of women and men. Gender roles are dynamic which change over time. In contrast, sex identifies the biological differences between men and women. The difference between gender and sex is indicated in Table 12.1.

Although women and men enjoy equal status in the law of the constitution, there exists a wide difference in how men and women are looked upon and handled by society. It is the society that has nurtured the 'Gender gap', as it is a socially constructed attribute of being male and female, by assigning roles and responsibilities with respect to one's gender.

In order to address gender issues, it is imperative that we understand the different fundamental gender concepts to deal with them effectively.

Table 12.1: Difference between Gender and Sex

Gender	Sex
Socially constructed	Biologically defined
Differs between and within culture	Determined by birth
Includes variables identifying differences in roles, responsibilities, needs, and constraints	Universal
Can be changed	Remains unchanged

Gender roles: Men and women have multiple roles. Women have productive, reproductive, and management of community assets roles whereas men have productive and other roles. Gender roles are learned behaviors in a given society/community, or other special group, that condition which activities, tasks, and responsibilities are perceived as male and female. Gender roles are affected by age, class, race, ethnicity, religion, and geographical, economic, and political environment. Some common terms relating to gender in use are indicated below.

Gender bias: The tendency to make decisions or take actions based on gender is gender bias.

Gender blindness: Gender blindness or sex blindness is the practice of disregarding gender as a significant factor in interactions between people.

Gender division of labor: It is the roles, responsibilities, and activities assigned to women and men based on gender.

Gender equality: Gender equality is the result of opportunities and the allocation of resources or benefits or access to services without any discrimination on the basis of a person's sex.

Gender equity: A condition in which women and men participate as equals, have equal access to resources, and have equal opportunities to exercise control and fair treatment for both women and men are given according to their respective needs is gender equity.

Gender awareness: The understanding of the socially determined differences between women & men based on learned behavior affecting their ability to access and control resources and benefits accruing therefrom is called gender awareness.

Gender discrimination: Differential treatment of an individual based on a gender stereotype is gender discrimination.

Gender mainstreaming: Gender mainstreaming is about considering gender in all policies, planning, budgets, and implementation and monitoring of programs/schemes, instead of addressing gender as a separate issue through separate programs. Gender mainstreaming is the process for ensuring that women and men have equal access and control over resources, development benefits, and decision-making, at all stages of the development process.

Gender sensitivity: It is the ability to acknowledge and highlight existing gender differences, issues, and inequalities and incorporate these into strategies and actions.

For example, gender sensitivity with reference to the above statements will be reflected if Govt. of India takes a decision to include the wife's name in land holding compulsorily, by law, to enable farmers to avail of benefits of schemes and programs.

Gender issues: Gender issues are the specific consequences of the inequality of women and men. It includes all aspects and concerns related to women's and men's lives and situations in society, the way they interrelate, their differences in access to and use of resources, their activities, and how they react to changes, interventions, and policies.

Gender relations: The ways in which a culture or a society defines rights, responsibilities, and identities of men and women in relation to one another is called gender relations.

Gender division of labor: The roles, responsibilities, and activities assigned to women and men, based on gender is called gender division of labor.

Gender empowerment: It is the process of generating and building capacities to exercise control over one's life.

Gender lens: The practice of investing for financial return along with considering the benefits to women, both through improving economic opportunities and social well-being for girls and women is a gender lens.

Gender needs: Since women and men have differing roles based on their gender, they will also have different gender needs.

Practical gender needs (PNGs): PGNs arise out of the gender division of labor and women's subordinate position in a society which are a response to an immediate and perceived necessities, identified within a specific context.

Strategic gender needs (SGNs): SGNs are related to the gender division of labor, power, and control, which include issues like legal rights, reproductive rights, domestic violence, equal wages, land ownership rights, etc. These may help women to achieve greater equality and change existing roles, thereby challenging women's subordinate positions.

Sex disaggregated data: Data required for gender analysis are segregated by sex/gender in order to allow differential impacts on men and women.

Gender analysis: Gender analysis is the most critical step forward towards gender-responsive planning and programming. It involves the collection and analysis of sex-disaggregated information and examines the differences, commonalities and interactions between women and men. Because both men and women perform different roles, they may have different experiences, knowledge, talents, and needs. Gender analysis explores these differences so that policies, programs, and projects can identify and meet the different needs of men and women.

Gender planning: Gender planning is a process of planning developmental programs and projects that are gender sensitive. It takes into account the impact of differing gender roles and gender needs of women and men in the target community or sector.

Gender budgeting: It is required for achieving gender mainstreaming and to ensure that the benefits of development reach women, as much as men. It is an ongoing process of including gender perspective in policy/program formulation, its implementation, and review. The three main components of Gender Budgeting are Policy Appraisal, Gender Budgeting, and Gender Auditing.

Women in Agriculture

With about 50% of the total population, women play a crucial role in achieving household food and nutritional security through agricultural production. They work as female agricultural laborers, as farmers, as co-farmers, as female family labor, and as managers of farms and farm entrepreneurs. Research on women's time utilization revealed that on an average women work for 15-16 hours a day out of which 7-8 hours in peak and 5-6 hours in lean season are spent in farm work. Although women currently provide 60-80% of the agricultural labor, they have limited access to the resources and opportunities needed to maximize and profit from their contributions. Participation of women in agriculture also varies in irrigated and rain-fed situations. Studies conducted in nine states in India, under All India Coordinated Research ProjectAICRP on home science of Directorate of Research on Women in Agriculture (DRWA) and (All India Coordinated Research Project (AICRP) Annual Report, 2007 & 08) indicated that independent participation of women was more in homestead-based agriculture. Joint participation in crop production activity was 75% for major crops, 79% for horticulture, and 51% in post-harvest operations. In livestock work participation rate of women was 58% and in fisheries about 95%.

Participation of Women in Agriculture

According to the Census of 2011, the sex ratio of India is 943 females per 1000 males. The national literacy rate is 74.04%, but the workforce participation rate for females is 25.51% against 53.26% for males. While considering participation in agriculture, 89.5% of the total female employed in rural areas, are engaged in agriculture and allied sectors and about 70% of farm work is performed by women. Women play a key role in the conservation of natural resources like land, water, flora, and fauna. There are specific agricultural operations like transplanting, weeding, harvesting, agro-processing, etc. which are done by women only.

Women in Animal Husbandry

In animal husbandry, women play multiple roles with regional differences. In the dairy sector, their roles range from cleaning of animal sheds, care of animals, milking, feeding, watering, grazing, fodder collection, chaffing and storage of fodder, cleaning of milking utensils, processing dung and milk into ghee, butter, *khoa,* etc. They also take health care of pregnant and sick animals and newborn or young calves. Sometimes they prepare cooking fuel by mixing dung with twigs and crop residues. Backyard poultry is considered as women-domain which serves family nutrition and income. The activities in goat farming performed by women are: caring for animals for breeding, care and management of newborn kids, collection & sale of manure, fodder collection, its cutting & chopping, taking animals for grazing, and marketing of animals.

Women in Sericulture

Out of 6.4 lakh villages in India, sericulture is being practiced in about 69,000 villages (Central Silk Board, 2002; Geetha and Indira, 2011; Lakshmanan et. al. 2011), engaging nearly 60% women of the total workforce in sericulture. In mulberry gardens, they work in the cultivation of silkworm food plants, inter-cultivation, weeding, application of Farm Yard Manure, leaf harvest & their transportation, pruning, and silkworm larvae rearing for production of raw silk. At post-cocoon technology, women are skillfully involved in silk reeling for unwinding of the very fine delicate silk filament, twisting, dyeing, weaving, printing, and finishing. At home also, women have the monopoly to look after silkworm rearing activities like leaf chopping, bed cleaning, feeding the silkworms, maintenance of hygiene, picking up the ripe worms and placing them on montages, and so on.

Women in Horticulture

Women play a significant role in horticulture including production, post-harvest operation, and value addition of fruit and vegetables. Participation of women in vegetable production is in field preparation, seed cleaning, seed sowing, transplanting of seedlings, weeding, harvesting, sorting, and grading of vegetables. Sometimes they also go for manure application. Women are trained for organic cultivation of selected vegetables, fruits, and medicinal plants. In fruit cultivation, some important operations like manuring, pot irrigation, harvesting, processing, and marketing are done by women. They generate highly remunerative employment through mushroom cultivation, vermi-compost preparation, floriculture, seed production, nursery maintenance, and tissue culture, seedling production of fruits & flowers, and food processing.

Women in Fishery

In India, out of the total 5.4 million active fishers, 3.8 million are fishermen and 1.6 million fisherwomen. According to FAO (2015), 70% of the global aquaculture workforce is female. In developing and developed countries, women are the bulk workers in postharvest, product transformation activities (FAO 2017). The Indian women are mostly involved in fish production, catching, cleaning, processing, peeling, drying, marketing, and net-mending. Sometimes they are also engaged in prawn culture, integrated fish farming, backyard hatcheries, fish-feed manufacturing, fish harvesting, packaging, value addition, net mending, and ornamental fish farming. In fresh and brackish water aquaculture, women are engaged in carp breeding and nursery raising, carp poly-culture, and breeding of catfish and freshwater prawns in backyard hatcheries. They are also engaged in preparing feeds and feeding fish, preparing ponds, stocking fingerlings, guarding the ponds during the daytime, pond maintenance, and harvesting. In the commercial fishery sector, women are involved as wage labor for de-heading, peeling, dressing, cleaning, washing, and icing. Through SHGs, women are engaged in the breeding and rearing of ornamental fishes, with specific species such as guppy, platy, molly, and sword tail. The women in small-scale aquaculture, in family ponds or tanks, help in providing protein, income, and employment.

Women in Forestry

In India, depending upon the socio-cultural variations, women are the major collectors and users of Non-timber Forest Products (NTFP). They collect fruits, nuts, roots, tubers, vegetables, mushrooms, saps, gums, medicines, seeds, household items, building materials, horns, eggs, honey, furs, wild leaves, and farm implements as they are aware of the properties and potential uses of these products. Women make specific contributions to forestry and agro-forestry value chains, particularly with respect to collecting, processing, and marketing of NTFPs for the food security of their households. In rural areas, women and girls fetch fuel wood from forest. In Odisha, plucking of Kendu leaf for *bidi* making and collection of Mahua flower for alcohol preparation are done by tribal women.

Women in Rural Production

Indian rural women are engaged in wage labor, non-farm, or off-farm income-generating activities. They are involved in livestock production, small enterprises, agriculture processing, and home gardens. They process NTFPs for the development of small-scale enterprises like a basket, broom, rope, and mat making, silk cocoon rearing, bamboo works, lac cultivation, oil extraction, leaf plate making, etc. Women work in the MGNREGA program for the enhancement of the quality of life of their families. The objective of these government programs/schemes is to reduce poverty and unemployment, improve health & educational status, and to fulfill the basic needs such as food, shelter, and clothing of the rural population.

Women in Food Security

At the household level women play an important role in ensuring food and nutrition security. They act as food producers of major grains and minor millet and processors of livestock and fishery products. Women are the decision-makers to provide the right food to the family members including pregnant & lactating mothers, old and sick persons, children, and others. They buy and prepare food for the family. According to Sabita et. al. (2005), farm women supply green leaves, vegetables& fruits, milk, and its products by rearing cows, and eggs through the rearing of poultry/duckery, local fish and crabs from local reservoirs, mushrooms through collection from the locality in the rainy season and honey through beekeeping. Food and nutritional security continue to be an important development agenda of the global community. This is clearly reflected in the Sustainable Development Goals (SDGs) of the United Nations Organization to be achieved by 2030.

Women in Natural Resource Management

Women contribute a lot in land use and its management. Indian women have played significant roles in several movements for the protection of forests such as the Chipko movement. They know how to conserve and maintain the quality of water. Women in hilly areas also have knowledge of animal ecology. They are more conversant with the resources around their homes.

Women in Home Management

Women play traditional roles as wives, mothers, friends, protectors, baby sitters, and so on. They collect fuel wood and drinking water for domestic consumption. They know the art of collecting, storing, reusing, and recycling local resources. It is the women who have more responsibility toward child rearing, preparing meals, cleaning, mopping, washing utensils & clothes, caring for the children, the elderly & the sick, collecting fuel, fodder & water, and caring for livestock. In addition, women are multi-tasked with income-generating activities and contribute significantly to household economic security. For this, they put in the constant effort and manage time with an advance planning, involving family members, and keeping things organized.

Considering the significant contribution of farm women in agriculture and the gender issues they are challenged with the programs/schemes, tools & implements, technologies, and research projects should be developed from women's perspective. Care should be taken to improve women's access and control over productive resources. The agricultural activities performed by farm women should be kept reserved only for them, to provide year-round employment to them. Empowerment of women should start from individual households, considering their mental and intellectual level.

Problems of Farm Women

Lack of Extension Service: All agricultural services still have a gender bias in favor of men. Women's work remains invisible to extension workers and policymakers so the extension and research priorities are directed to cash and export crops rather than food crops for domestic consumption. Hence much research and extension support is not available on technologies appropriate for women's multiple tasks. It is often assumed that men are the heads of households and they will pass on the extension information to women in the households. The male migration to distant places in search of work has led to increased women-headed households but enough technological support is not available to such women. The triple burden of women in child rearing, farm work, and household responsibilities are not given due consideration during designing a training program. Its venue, timing, duration, content, and methodology are not very appropriate for women.

Gender issues in land ownership: Women's lack of ownership of land/ access to land continues to be a major obstacle to increasing their contributions and benefits. Limitations of land access hamper the long-term planning for women. Women-headed households suffer from land constraints as their credibility to resources also gets affected. Women who do not own the land they work on are less inclined to invest precious time and scarce resources in long-term and improvements such as irrigation or drainage systems, terracing, tree planting, and other activities that maintain soil fertility. Moreover, women who do not own land are usually denied access to agricultural support services such as credit for purchasing inputs, training in land and water development, and water supplies for irrigation.

Over burden of work: Women suffer the brunt of triple responsibilities (agricultural production, reproduction, and nurturing) and they are involved in labor intensive, monotonous, repetitive, and drudgery-prone work, which is carried out manually, leading to mental and physical exhaustion and occupational health hazards.

Gender bias in development policies: Since women's work is invisible & their contribution is not recognized as paid work they are treated as "consumers" rather than "producers". Hence development policies are lopsided not always favoring women.

Limited access to production resources: Farm women play crucial roles in the production, storage, and processing of food in most societies. Despite playing a pivotal role in food production women face several hardships, in case of the breakdown of a family due to the death of a husband, divorce, or desertion; women are rendered landless because no assets are no women's name. Even if they have land they are constrained for money and other resources (inputs and technical know-how) required for cultivation. Agricultural development programs are usually planned by men and aimed at men.

Out-migration of men: In many rural areas, migration of men and other changes in farming systems are placing even greater burdens on women, who are left behind to manage agriculture and the entire household alone.

Socio-economic Status: Women's share of work in agriculture is greatest among small farmers. When holdings are larger, women may withdraw from work in the field or supervise hired female labor. Women in landless households spend twice as much time working for wages in agriculture than do women in families with land.

Shift towards cash crops: The introduction of cash crops changes the pattern of household labor allocation by diverting labor from subsistence crops. When women are expected to provide labor for cash crops, they have less time available for subsistence agriculture. The income from cash crops usually comes under the immediate control of men through the practice of directing inputs, credit, and extension to heads of household.

Limited access to market: Women's efforts to expand the volume of their income-generating activities are thwarted by their limited access to marketing facilities and services. Although women worldwide are active as traders, hawkers, and street and market vendors, little has been done to assist them with improved transport and market facilities.

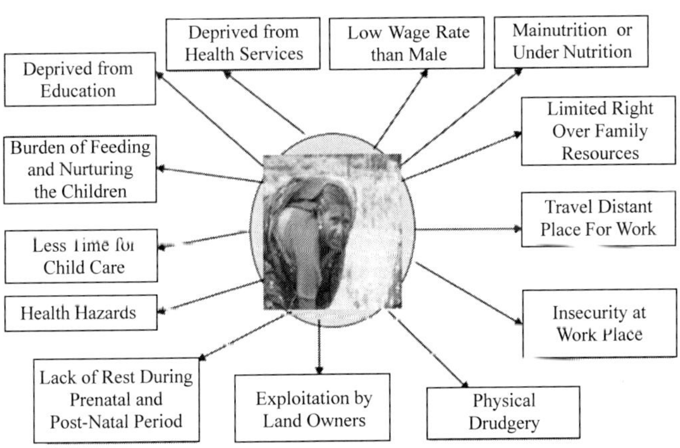

Figure 12.1: Major issues of rural women in agriculture.

Seasonal employment: Women bear the brunt of hardship arising out of seasonal unemployment/underemployment.

Lack of education: Lack of education forces women to work as unskilled labor with lower wages. One study on agricultural productivity showed that four years of primary education increased farmers' productivity by up to 10%, and the benefits of education for women farmers can be even greater.

The important issues of women in agriculture are indicated in Figure 12.1.

Gender and Agricultural Development

Agriculture involves both self-employment and wage-employment and accordingly requires precise information about who does what. In fact, the land holding of 75% of the farming community being small, the number of landless laborers has swelled up over time by working on others' farms. The situation thus demands an understanding of the activity performance of men and women; and the children – girls and boys, whose lives are fundamentally structured in different ways. Their living pattern, work patterns, interaction styles, and sharing of scientific information differ within the socio-economic groups.

Similarly, a gender-based division of labor is universal but cultural and community diversities cause differentiation.

Recognizing the social characteristics of gender, and building gender into agricultural development will lead to:

- Building inherent strength of women & men to mutually learn and share.
- Overcoming gender distorted prejudices.
- Articulating gender perspectives in development activities.
- Involving action from women to men and from men to women for gender sensitization.

Gender Budgeting

The conventional approach to gender budgeting, i.e., isolating public expenditure- direct and indirect – for women, would continue to be a core activity under the broad gender budgeting exercise with future action concentrating on fine-tuning methodology and universalizing the tools for application at all levels of public expenditure. Gender budgeting does not seek to create a separate budget but seeks affirmative action to address the specific needs of women.

Why Gender Budgeting?

National budgets reflect how governments mobilize and allocate public resources, and how they aim to meet the social and economic needs of their people. The rationale of gender budgeting arises from the recognition of the fact that national budgets impact various sections of society differently through the pattern of resource allocation and priority accorded to competing sectors. The purpose of the gender budgeting exercise is to assess the quantum and adequacy of the allocation of resources for women and to establish the extent to which gender commitments are translated into budgetary commitments. This exercise facilitates an increase in accountability, transparency, and participation of the community. The macro policies of the Government can have a significant impact on gender gaps in various macro indicators related to health, education, income, etc. Gender mainstreaming requires gender-responsive policy.

Evolution of the Concept of Gender Budgeting

The perspective on gender budgeting has greatly evolved over the last few years from the initial *post facto*, statistical exercise that sought to establish a quantum of resources allocated for women in the Union and State Budgets. The range of gender budget analysis is now perceived as a mandate to examine with a gender perspective.

Gender Mainstreaming – The New Mantra

Gender mainstreaming intends to achieve gender equality and improve the relevance of development agendas. The ultimate goal of gender mainstreaming is to achieve gender equality and equity. From the perspective of gender budgeting, the public expenditure in social sectors like education, health, and rural development should specify the allocation for women. It is necessary to recognize that women are equal players in the economy whether they participate directly as workers or indirectly as members of the care economy. To that extent, every policy of Government fiscal, monetary, or trade has a direct impact on the well-being of women. Thus, it is not adequate to analyze in detail, the allocation of resources for a few sectors of the economy that are traditionally considered as women related. It has to embrace a gender-sensitive analysis of monetary policies, covering the impact of indicators like inflation, interest rates, etc., and fiscal policies covering taxation, excise, etc. Thus, gender budgeting analysis has to go hand in hand with gender mainstreaming. This is a mammoth task in the current scenario when there is a lot of disaggregated data that has to be gender sensitized and analyzed. Micro studies will have to be conducted to study the impact of public and fiscal policy so that they may guide macro policies. Thus, a wide gamut of activities needs to be undertaken for gender mainstreaming.

There was provision definite flow of funds from the general developmental sectors to women in the Eighth Plan with the intention that the benefits to development from different sectors should not bypass women and special programs on women should complement the general development programs.

The Ninth Plan (1997–2002) adopted the "Women Component Plan' as one of the major strategies and directed both the Central and State Governments to ensure "not less than 30 per cent of the funds/benefits are earmarked in all the women's related sectors." The National Policy for Empowerment of Women 2001 envisaged the introduction of a gender perspective in the budgeting process as an operational strategy. The Tenth Plan reinforced commitment to gender budgeting to establish its gender-differential impact and to translate gender commitments into budgetary commitments. The Approach Paper to the Eleventh Five Year Plan mentioned "Gender Equity requires adequate provisions to be made in policies and schemes across Ministries and Departments."

Strategic Framework of Gender Budgeting Activities

After a consultation process, the Ministry of Women and Child Development has adopted "Budgeting for Gender Equity" as the Mission Statement for this initiative and has drawn up a broad framework of activities that would constitute action areas for gender mainstreaming. Quantification of allocation of resources for women in the Union, States, and Local Administration budgets and expenditure thereof should consider:

- Refining and standardizing methodology and development of tools.
- Trend analysis.
- Analysis of change in pattern, shift in priorities in allocation across clusters of services, etc.
- Variations in the allocation of resources and actual expenditure.
- Adherence to physical targets.

Strategies for Mainstreaming Gender

- **Organizing Women Groups**: Male extension workers can be trained to work more closely with women in settings that are culturally acceptable, such as women's groups.
- **Technologies to reduce energy and time spent, particularly the household and farm production activities**: Women-friendly technological innovations such as weeders, paddy threshers, winnowers, sprayers, harvesting tools, parboiling units, maize shellers, dal-making machines, etc. will be promoted to reduce women's drudgery.
- **Increasing the biomass production to meet fuel needs**: Plantation of fast-growing fodder in common lands and developing mechanisms for sharing the fodder helps women in saving a lot of time and devote this time to income-generating activities.

- **Credit**: Innovative credit programs using non-traditional forms of collateral and local institutions (women groups) can ensure that women are able to obtain access to credit.
- **Capacity building**: Identifying the right training and extension needs of women is one of the most important steps in initiating any developmental program. Gender analysis of activities, resources, constraints, implications, and benefits should be understood using a participatory approach. This information should be taken into consideration for the need assessment. The staff members' / extension functionaries' ability to do this has to be built up. Farm women should be given more access to meetings, training, exposure visits, and demonstrations, organizing training programs based on the needs of the women. Institutional and village-based training to be organized as per the convenience of the women farmers. Active women can be selected and trained, and they should be provided with inputs and credit to practice improved technology.
- **Collective farming**: Where severe fragmentation exists, collective farming should be encouraged by women.
- **Extension service**: More women extension workers should be recruited from rural areas and trained. Female para-extension agriculturists, relatively uneducated women with short crash courses on agriculture, can be posted in their own villages. Female supervisors and subject matter specialists (SMSs) need to be posted. In the case of male extension workers, the stereo-typed attitude of male agents should be changed with regular gender sensitization courses. They should be trained in increasing awareness of gender roles and developing skills in the use of language and communication to suit women.
- **Health care**: Crèches for children of women farmers should be provided at work sites. Proper healthcare support is needed for girls and women.
- **Marketing support**: Most of the micro-enterprises undertaken by women are based on the skills and raw materials available rather than considering the market needs and market dynamics. An in-depth marketing study would help identify effective marketing strategies for products. While developing farm women programs, the cost of hiring consultants to conduct market analysis and market development should be kept in mind.

Agricultural Programs for Women

The DAC & FW, Ministry of Agriculture & Farmers Welfare, Govt. of India has made special provision for women in many programs and schemes

National Mission on Agricultural Extension & Technology (NMAET) and its four sub-missions. Besides the following programs are being implemented for the development of women in agriculture.

- Support to Training & Employment Program for Women (STEP) in 1986-1987.
- Rashtriya Mahila Kosh (RMK) in 1993.
- Gender Budgeting Scheme (GBS) from 2004 to 2005.
- National Gender Resource Centre in Agriculture (NGRCA) from 2005 to 2006.
- Mass Media Support for Agriculture Extension.
- Mahila Kisan Sashaktikaran Pariyojana (MKSP) from 2010 to 2011.
- Mahila E-haat in 2016.

References

Central Silk Board. 2008. Annual report, 2007-08. Bangalore, India: FAO (2015): Women in Eastern and Central Asia.

Geetha, G. S., & Indira, R. 2011. Silkworm rearing by rural women in Karnataka: A path to empowerment. Indian Journal of Gender Studies, 18, 89-102.

National Research Centre for Women in Agriculture (2008): Annual Report.

Mishra, S 1991. Role of Television in diffusion of Home making practices among urban housewives, Ph.D. thesis, Utkal University, Vani Vihar, Bhubaneswar.

Suggested Reading

Farm Women Friendly Hand Book: A Handbook for Women Farmers 2018. National Gender Resource Centre in Agriculture Ministry of Agriculture & Farmers Welfare Government of India.

Gender and Food Security: Published by FAO.

Gender Budgeting Handbook for Govt. of India, Ministries and Departments 2007. Ministry of Women and Child Development, Govt. of India.

GOI 2009. Gendering Human Development Indices: Recasting the Gender Development Index and Gender Empowerment Measure for India. New Delhi: Ministry of Women and Child Development, Govt of India with UNDP India. https://www.undp.org/content/dam/india/docs/gendering_human_development_indices_summary_report.p.

Grover, Indu. 2003. Unit-I Gender Theory and Development: Post Graduate Diploma on Gender and Development through Distance Education, HEL Project,. Hisar, India: Department of HSEE, CCS HAU. Hisar, India.

Guidelines for Gender analysis of Public Expenditure: Ministry of Women and Child Development, Govt. of India.

MANAGE, Hyderabad 2021. Gender Mainstreaming for Agricultural Development.

Singh S (2017) The State of Gender Inequality in India. Gender Stud 15(1):139–157. https://doi.org/10.1515/genst-2017-0009.

UNESCO Gender Mainstreaming Implementation Framework (GMIF), 2002-07. the section for Women and Gender Equality of the Bureau of Strategic Planning, Paris, France.

Index